Lecture Notes in Earth System Sciences 144

For further volumes:
http://www.springer.com/series/10529

Jonathan David Istok

Push-Pull Tests for Site Characterization

Jonathan David Istok
School of Civil and Construction Engineering
Oregon State University
Corvallis
97331 Oregon
USA

ISSN 2193-8571 ISSN 2193-858X (electronic)
ISBN 978-3-642-13919-2 ISBN 978-3-642-13920-8 (eBook)
DOI 10.1007/978-3-642-13920-8
Springer Heidelberg New York Dordrecht London

Library of Congress Control Number: 2012945899

Printed on acid-free paper

Springer is part of Springer Science+Business Media (www.springer.com)

Push-Pull Tests for Site Characterization

Jonathan ("Jack") Istok, PhD, PE
Professor
School of Civil and Construction Engineering
Oregon State University
Corvallis, OR 97331
Jack.Istok@oregonstate.edu
541-737-8547

Summary

This book has described and hopefully demonstrated that the single-well, push-pull test, in all its various manifestations, is a useful tool for site characterization, pilot testing of various remediation strategies, and so on. The advantages of in situ testing are clear, and the push-pull format makes it possible to perform field experiments at relevant scales for a modest cost and with no more logistical complexity than a laboratory experiment. Often in the biogeochemical sciences, we think of research as a linear process from small-scale laboratory development to ultimate (usually at a date a long way off) full-scale field application. However, using the push-pull test it is possible to conduct field experiments in parallel with laboratory experiments and this should lead to better synergy between the two. Simple field experiments have the remarkable ability to sharpen experimental design by identifying the most important parameters to control in laboratory experiments and to quickly select the most promising small-scale procedures and techniques to further develop in the laboratory. Nevertheless, much developmental work on the push-pull test format remains to be done. Additional numerical and analytical methods for interpreting push-pull test breakthrough curves for the ever more complicated biogeochemical systems being studied are needed. The design of push-pull tests can be refined to incorporate valuable auxiliary data such as isotopic composition of test solutions, microbial community dynamics, surface chemistry reactions, and so on. To date, only a limited number of push-pull tests have been conducted (using gas phase tracers) in the vadose zone and these have focused on a single microbial process (methane oxidation). Most push-pull tests have been conducted in the near surface (<100 m deep), but the test is potentially most valuable in deep subsurface environments where the costs of drilling for sample collection are very high. Thus, push-pull tests in deep boreholes on land and beneath the seafloor hold great potential in elucidating important biogeochemical processes in those environments. Combining push-pull tests with various geophysical imaging techniques also holds great promise because of the ability to modify the subsurface environment (e.g., by injecting electrolytes that can be detected by electrical conductivity arrays) in defined ways and thus identify preferential flow paths, reactive surface areas, etc.

Contents

Chapter 1
Introduction

1.1 The Need for In Situ Testing

There is a continuing need for quantitative information about the subsurface environment. Applications include developing and managing new or existing water supplies, storm water management, energy storage and recovery, environmental remediation, artificial recharge, gas and oil production, carbon sequestration, and many others. The spectrum of processes that might be of interest include essentially all the physical, chemical, and biological processes that are operational in any natural or engineered environment. Physical processes might include fluid flow by advection and mixing/spreading processes such as dispersion, conduction, and convection that apply to transport of matter and energy. Chemical processes might include the wide range of chemical reactions that are possible in water/mineral/atmosphere systems such as acid/base, precipitation/dissolution, sorption, surface complexation, and oxidation/reduction reactions. Biological processes might include any of the diverse metabolic activities displayed by indigenous microorganisms as they obtain energy and nutrients for growth and include processes related to nutrient cycling, detoxification of contaminants, and many others. Of course all of these processes are coupled so that physical properties will influence fluid flow, which controls the flow of substrates that in turn influence microbial growth, or solutes that participate in chemical reactions. There is no satisfactory term to describe the coupling among all these processes; "biogeochemical processes" will be used here with the explicit understanding that physical processes are also included.

Identifying and characterizing all biogeochemical processes pertinent to a particular problem is challenging in any system but is typically much more so in subsurface environments. Obtaining access to the subsurface for direct observations or measurements can be difficult and expensive. Although indirect methods (e.g., ground-penetrating radar, seismic and electrical surveys, etc.) can provide much useful (although typically somewhat qualitative) information on the physical characteristics and some processes, direct sampling and laboratory testing remains

J.D. Istok, *Push-Pull Tests for Site Characterization*, Lecture Notes in Earth System Sciences 144, DOI 10.1007/978-3-642-13920-8_1,

the principal method for obtaining quantitative information about biogeochemical processes in subsurface environments. Gas and liquid samples are routinely collected from wells, boreholes, lysimeters, etc. but provide only a partial description of the subsurface environment because, for example, they may exclude reactive mineral phases, attached microorganisms, etc. Pressure changes and exposure to the atmosphere within a well or during sampling can release dissolved gases such as carbon dioxide (raising pH), hydrogen, and methane from pore fluids. Many subsurface environments are anaerobic and/or chemically reducing and exposure to the atmosphere during sampling may introduce atmospheric oxygen that disrupts microbial processes or reacts with reduced aqueous species in the sample.

Although sediment samples are routinely collected by various coring methods the samples retrieved are typically very small (a few tens to a few hundreds cm^3) and therefore are almost certainly not representative of the larger-scale heterogeneities that are present in all natural environments. Moreover, such samples are subject to a variety of disturbances during the collection process (heating, drying, wetting, compression, contamination with drilling fluids, etc.) that can alter sample composition or introduce artifacts that compromise laboratory measurements. For example, particles larger than a few cm may be too big to enter core tubes and certain other particle sizes (e.g. fine sands) may be lost when these devices are extracted from the borehole. Pore space and geometry may be changed by compression or expansion as sediments enter core barrels or are transferred to laboratory apparatus for testing. It can be very difficult to collect intact samples in some materials (e.g., fractured rock or loose sands) and it is not uncommon to recover only 10–25 % of the intended sample from a core tube. Sediment samples can also be very difficult and expensive to obtain, especially at depths greater than ~10 m, in materials that are difficult to penetrate (e.g. hard rock), in highly contaminated environments, or in locations near or beneath buildings or other infrastructure. Drilling costs in such environments can also be very high, severely limiting the number or size of samples that can be collected. In very deep environments it may only be possible to collect samples from a single well or borehole. Although it is sometimes possible to collect cores from within a well (e.g., by angled-drilling through the well casing), generally sediment samples can only be collected once at each location with additional samples requiring additional boreholes. Thus, at many sites there are a large number of existing wells, installed for previous site characterization campaigns and there is a general “reluctance” by site owners to install more wells to obtain “fresh” sediment samples for laboratory testing. Moreover, in many cases regulators will require that a borehole created to collect sediment samples, be completed as a conventional monitoring well, thus incurring the high costs of routine monitoring for an indefinite time into the future.

In conventional practice, biogeochemical processes of interest are identified and quantified in laboratory experiments using sediment and or groundwater samples. For example, almost all existing information on biogeochemical processes in subsurface environments has been obtained by placing sediment and/or groundwater samples into closed vessels (“microcosms”, columns, etc.) and monitoring reaction progress by periodic aqueous or sediment sampling. However, it is well known that this approach introduces various "artifacts" that can greatly influence the results and conclusions from such experiments. For example, for convenience, temperatures, pressures, and sediment:fluid ratios employed in laboratory measurements or experiments with subsurface material are often different from in situ conditions, and drying/sieving/shaking/homogenization are also typically employed to create uniform materials for various experiments. All these operations tend to change the particle size distribution, inter-particle pore geometry, and surface area of sediments, which can greatly influence their reactivity. For example, crushing mineral grains exposes reactive surfaces that are chemically reduced and will react with and consume oxidants, whereas the surfaces of uncrushed mineral grains are typically already oxidized and therefore less reactive. Many mineral surfaces are irreversibly altered by drying, which also changes their reactivity. Many experiments employ vigorous mixing to obtain uniform composition within the reactor, which of course never occurs in subsurface environments where laminar flow limits "mixing" to the very weak processes of diffusion and dispersion. Vigorous mixing generally increases the rates of biogeochemical processes and it is not uncommon for laboratory and field reaction rates to vary by many orders of magnitude.

It is often inconvenient to use site groundwater in laboratory experiments and often a "synthetic groundwater" is prepared from distilled water with reagent grade salts added to match the major ion composition of site groundwater. However, a truly identical composition is almost never achieved because of trace constituents that are not identified or that are difficult to include in the synthetic groundwater recipe (e.g. hydrogen) or maintain over the duration of the experiment (e.g. pH or dissolved carbon dioxide). Moreover, some in situ processes are difficult to study in the laboratory for health or safety reasons, e.g. if reactions of interest involve high concentrations of solvents, radionuclides, or other hazardous materials. Costs of disposing of sediments contaminated with these materials can also be high and serve to limit the size and number of laboratory experiments that may be performed using samples from such environments.

Although less is known about the effects of drilling, coring, and subsequent laboratory manipulations on the composition of the microbial community or the activity of particular microbial groups, it can be inferred from other systems there will likely be changes in microbial physiology due to sample handling. For example, the introduction of oxygen may disrupt the activity of obligate anaerobes (especially when pore structure is disrupted by sample homogenization), many of which are important for pollutant detoxification. Characterization of microbial community composition is typically performed using molecular techniques and it is quite likely that sample handling will have important effects on e.g. the efficiency of biomarker recovery. It has been shown, for example, that simply enclosing a seawater sample in a glass bottle changes the microbial community composition, which is sometimes called the "bottle effect". Similar "bottle effects" are also expected in laboratory microcosms and columns. Rates of chemical and biological reactions seem particularly sensitive to a wide range of laboratory artifacts and disturbances. Reaction rates typically depend on concentrations of all reactants, temperature, pressure, and other factors. In closed systems, metabolic produces, which would normally be dispersed by diffusion, can accumulate, reducing the free energy yield of targeted reactions. Laboratory rate determinations are often made using higher temperatures than in situ conditions to avoid the need for conducting experiments in constant temperature rooms or incubators, or to deliberately speed the rate of slow reactions, but higher temperatures also affect solubilities, vapor pressures, etc. and the combined effects can be complex. Similarly, reactants may be introduced at higher concentrations in laboratory experiments than exist in situ to allow use of less sensitive and lower cost analytical equipment, again with a wide variety of potential unintended consequences (e.g. from increased ionic strength or mineral precipitation). Given all these factors, it is not surprising that laboratory rate measurements often differ from in situ rates by three or more orders of magnitude.

There is clear and abundant evidence that the specific details of sample collection and laboratory testing methodology can profoundly affect the results and these should always be carefully evaluated before any observations or measurements made in the laboratory are used to predict in situ behavior. This has long been recognized in hydrogeology where, for example, laboratory measurements of hydraulic conductivity are considered greatly inferior to in situ measurements made by pumping tests or slug tests. Similarly, in situ strength testing is recognized as superior to laboratory measurements of strength properties on sediment cores. For all the reasons cited above, most environmental professionals would agree that in situ testing of biogeochemical processes is superior to laboratory testing.

1.2 What Is a Push-Pull Test?

This book describes a general purpose field technique for quantitatively characterizing biogeochemical processes in subsurface environments. The technique is called a *"push-pull" test*. Push-pull tests can provide quantitative information on a variety of subsurface characteristics but are particularly useful for measuring in situ rates of chemical or microbial reactions. These rates are needed, for example, for use in reactive transport modeling and are often critical factors in environmental decision making. Push-pull tests are conducted entirely in the field using any device that provides access to the subsurface including conventional monitoring wells, drive-point wells, multi-level sampling wells (i.e. wells constructed with many sampling ports vertically isolated from each other), open boreholes, and similar devices. A typical test involves the injection ("push") of a prepared *test solution* into the subsurface at a single location (e.g. the well screen of a conventional monitoring well or the sampling port of a multi-level sampling well) followed by the extraction ("pull") of the test solution/pore fluid mixture from the same location. In the saturated zone, where pores are mostly filled with water, the test solution is an aqueous mixture. In the unsaturated or vadose zone, where pores are mostly filled with gas, the test solution is a gas mixture. In both cases, the test solution typically contains *nonreactive tracers* to assess the rate and extent of dilution of the injected test solution with ambient pore fluids and one or more *reactive tracers* to target particular biogeochemical processes. We will use the word 'tracer' to refer to any constituent whose concentration is of interest, whether already present in the ambient pore fluid or specifically added to the injected test solution by the investigator. The type, combination, and concentration of nonreactive and reactive tracers are selected to provide information on the processes of interest. Thus, nonreactive tracers can provide information on the physical processes of advection, dispersion, diffusion, etc., while reactive tracers can provide information on sorption, cation exchange, biogeochemical reaction rates, and so on. The historical development of the push-pull test and a wide range of specific applications are presented in subsequent sections.

The volume of injected test solution is selected to interrogate the desired volume of the subsurface environment. During the *injection phase*, the test solution is introduced into the subsurface where it penetrates an irregularly shaped volume, roughly centered about the injection location. Because of inevitable heterogeneities in properties that affect fluid flow (e.g. porosity and permeability) the shape of the interrogated zone is typically not precisely known. However, differences in the electrical conductivity of injected test solution and ambient pore fluids have allowed some investigators to use surface and downhole geophysical techniques to "image" (via tomography) the interrogated volume. During and after injection, the test solution drifts with regional flow away from the injection location and is diluted by the ambient pore fluids through the usual processes of advection, dispersion, diffusion, etc. During the *extraction* or *sampling phase* of the test, samples of the injected test solution/pore fluid mixture are collected at the injection location and concentrations of nonreactive and reactive tracers as well as potential reaction products that may have formed in situ are measured. The extraction phase may consist of either continuous pumping with periodic sampling or intermittent pumping and sampling, depending on the purpose of the test (Fig. 1.1). In general, sampling continues until tracer concentrations are too small to be useful in quantifying the targeted processes. For example, reaction rates are typically computed from plots of dilution-adjusted concentrations of reactive tracer consumed and/or reaction products formed and require measurements of nonreactive tracers to compute the necessary dilution factors. In this type of test, sampling might continue until nonreactive tracer concentrations fell below analytical detection limits.

By varying test conditions (primarily test solution composition and volume, sampling frequency, and sample analyses) it is possible to investigate a wide range of subsurface processes as described in the following examples. However, the push-pull test is particularly well-suited for measuring reaction rates and the majority of all tests have been performed to investigate rates of chemical or microbiological reactions in groundwater aquifers, particularly with respect to contaminant transformations that are important for waste site characterization and remedial design.

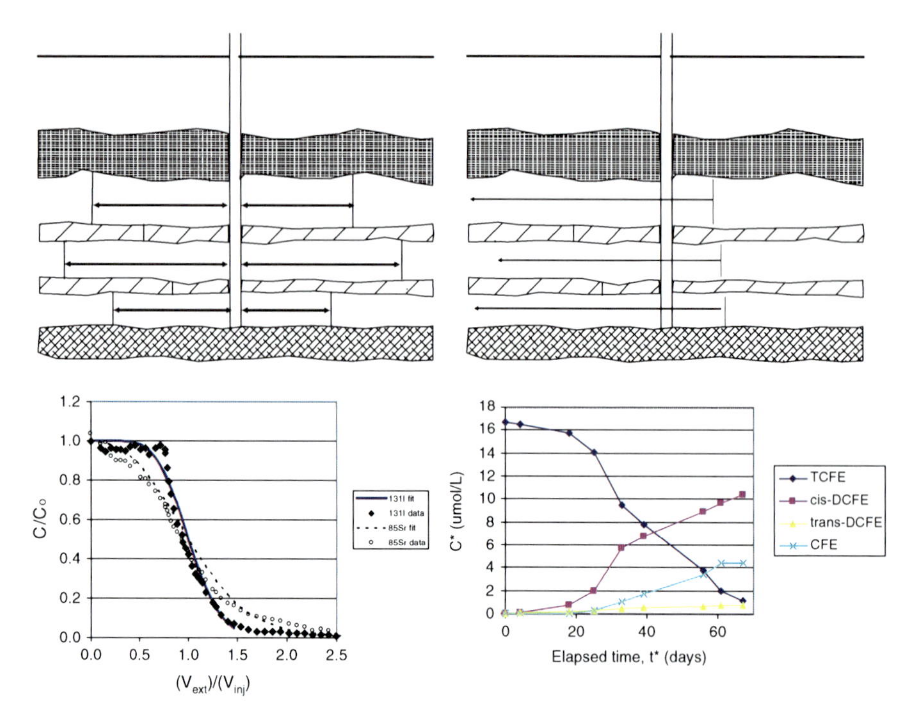

Fig. 1.1 Two types of push-pull tests with example data. (*left*) Test solution injection followed by continuous extraction pumping and sampling (*right*) Test solution injection followed by periodic discrete sampling. *Arrows* in *upper* figures show idealized test solution flow paths. Figure on *left* shows effect of sorption on concentration profiles for injected ^{131}I (nonsorbing tracer) and ^{85}Sr (sorbing tracer); *lines* show model fits. Figure on *right* shows in situ transformation of injected TCFE to cis and trans-DCFE and CFE. The rationale for these tests and the definitions of all acronyms and symbols are in the cited examples

1.3 Push-Pull Test Design

Push-pull test design involves: (1) identifying the locations to be tested, (2) selecting the volume of test solution to be injected, (3) specifying the test solution composition, especially with respect to the number and kind of nonreactive and reactive tracers to be used, and (4) selecting the extraction sampling frequency and number and types of sample analyses.

Any opening that provides access to the subsurface may potentially be used for push-pull testing. These may include the screened intervals of traditional monitoring wells, open boreholes, temporary or permanent drive-point wells, multi-level monitoring wells, suction lysimeters, etc. Tests have been conducted in the vadose zones and saturated zones of terrestrial subsurface environments and in saturated sediments beneath lakes, bays, and estuaries. Syringes and needles have been used to perform small-scale push-pull tests in the upper few centimeters of saturated soils (e.g. in studies of nutrient cycling in wetlands); wells, pumps and tanker trucks have been used to perform very-large scale, push-pull tests in deep groundwater aquifers. Tests have been conducted in the open water columns of lakes and other surface water bodies. In this type of application, care must be taken to avoid rapid dilution of injected test solutions (e.g. by wind driven mixing of the water column, or by high injection rates, which can create turbulence at the injection point), which leads to rapid reduction in tracer concentrations.

The ability to avoid new drilling and well construction and to use existing wells is a major advantage of the push-pull test and can greatly reduce the costs associated with in situ testing. Thousands of wells and similar devices have already been installed at hundreds of sites in many diverse environments and represent an underutilized resource for scientific investigation, a vast potential "underground laboratory" for learning about the subsurface environment without the cost of installing additional boreholes to obtain the "fresh" sediment samples typically required for laboratory testing. Some examples include geotechnical test wells, environmental monitoring wells, domestic and municipal water supply wells, irrigation wells, energy and mineral exploration or production wells, storm water, waste water, or liquid waste injection wells and so one. Existing wells often have some basic information available including descriptions of the materials and conditions encountered during well installation, pumping or slug test results, and perhaps historical water level and chemical composition data, which can be helpful when selecting specific wells for testing and may provide other information useful for successful test design or for data interpretation. For example, the permeability of the formation is always an important factor in test design because it may limit injection/extraction rates, which will influence test duration and cost (although this is a less important factor when injecting gaseous test solutions than when injecting aqueous test solutions). In saturated zone tests, the regional groundwater velocity (e.g. the seepage or pore water velocity) is also important as it controls the rate of dilution of injected aqueous test solutions (in vadose zone tests, Fickian diffusion typically controls the rate of dilution of injected gaseous test solutions). Perhaps, the most difficult test to implement is one designed to measure the rate of a very slow reaction or process when the regional groundwater flow velocity or diffusion rate is very large. In these cases, injected test solutions may be swept away from the test location before useful information can be obtained. Nevertheless, tests have been successfully conducted in thousands of wells in a wide range of geologic materials from fractured rocks to clays and have detected and quantified reactions with half-lives of minutes to months.

Figure 1.2 shows a few typical devices used in push-pull testing. In conventional monitoring wells, test solutions are injected and extracted through the well screens using pumps. In shallow submerged or saturated sediments, large-bore syringe needles or small-diameter well screens ("drive points") can be manually inserted to the desired depth and test solutions injected and extracted using syringes or small hand pumps. Push-pull tests can also be conducted within specific depth intervals by injecting and extracting test solutions into wells screened at different depths, using multi-level sampling wells or by using "straddle" packers to isolate a portion

Fig. 1.2 Push-pull tests may be conducted (clockwise from *top center*) using conventional monitoring wells, syringes and needles and temporary well screens in submerged sediments, in multi-level samplers, in drivel-point wells and in wells equipped with "straddle packers" to isolate a portion of the well screen to allow depth-specific testing

of the well screen. Additional details of field equipment are presented in the examples discussed in a subsequent section.

Selecting the volume of test solution to inject depends on several factors. The well and associated tubing "dead volume" (i.e. the volume of fluid stored in the well casing and access tubing) places a lower limit on the volume of injected test solution (which is why smaller diameter wells are usually preferable for testing compared to larger diameter wells, all other factors being equal). Obviously, a sufficiently large test solution volume must be injected to insure that the injected test solution actually penetrates into the targeted geologic formation (unless the test is specifically designed to interrogate only the in-well or in-borehole environment). Beyond this minimum requirement, the test solution volume is selected to be (1) large enough to interrogate a site- and problem-specific "representative elementary volume", (2) large enough that the rate or extent of dilution of the injected test solution will not adversely affect test results, and (3) sufficiently small that test logistics are not unnecessarily complicated. Hydrogeologists have long used the term "representative elementary volume" to refer to a hypothetical volume of subsurface material that is large enough to provide useful information about a targeted process given the inevitable heterogeneity present in all natural environments (e.g. spatially variability in the morphology of sediment and rock layers; the size, orientation, and density of fractures; sediment particle sizes, mineralogy, microbial and geochemical composition; etc.). The selection of a suitable representative elementary volume depends on a number of site specific factors but is ultimately subjective. For reference, many hydrogeologists would likely agree that ~1 m^3 (~1,000 L) would be considered a reasonable representative elementary volume at many sites. Smaller test solution volumes may be desirable in homogeneous materials or when it is desirable to examine small-scale heterogeneities in system properties (e.g. by performing a series of identical tests as a function of depth). Although larger test solution volumes are often desirable to obtain a larger sample in very heterogeneous environments or to slow dilution rates when ambient pore fluid velocities are high they may be difficult or costly to prepare or injection times may be excessively large. Thus, selecting a test solution volume is a process of balancing all these factors. It should be noted that tests have been successfully conducted with test solution volumes ranging from 10^{-5} to 10^{5} m^{3} but the majority of all tests have employed test solution volumes of 100–400 L (0.001–0.004 m^{3}). It is important to note that these volumes are many times larger than the typically much smaller volumes used when in laboratory experiments. Much of the existing information on subsurface biogeochemical processes has been obtained from laboratory experimentation on samples of only a few tens of cm^{3} and are therefore substantially biased by sampling artifacts. Important logistical factors in selecting the volume of injected test solutions include (1) the time required to inject and extract large volumes of fluids from the subsurface, especially in low permeability materials (e.g. fine clays or dense rocks), (2) the cost of suitable containers, reagents, tracers, etc. required to prepare large test solution volumes, and (3) the time and energy required for mixing such large volumes, and (4) waste disposal costs or costs of compliance with applicable health, safety, and environmental regulations.

Perhaps the most critical and scientifically interesting part of a push-pull test design is the specification of the composition of the injected test solution because the changes in test solution composition during the test provide the basic data that are used to detect, quantify, and interpret targeted biogeochemical processes. Test solutions may be prepared from ambient pore fluids (e.g., pore water or gases pumped from the subsurface prior to the start of the test) or from synthetic pore fluids prepared for this purpose, depending on the objectives of the test. The test solution is typically amended with one or more tracers or has its composition modified in some other way. Nonreactive tracers such as Cl^{-} or Br^{-} (for aqueous test solutions) and SF_6 or He (for gaseous test solutions) are almost always added to quantify dilution losses that occur as the injected test solution gradually drifts away from the injection location. It should be noted that "naturally occurring" nonreactive tracers are often present in ambient pore fluids and they can also be used to quantify dilution, sorption, and mass transfer processes as long as there is a detectable difference in the "chemical signature" between the injected test solution and the ambient pore fluids. These may include differences in measured concentrations of targeted anions, cations, dissolved gases, isotopic composition, etc. or other parameters including pH, electrical conductivity, or temperature. If suitable naturally occurring tracers are present it is sometimes possible to obtain the desired information by simply injecting tracer-free fluids (air or distilled or tap water), which can be extremely desirable in contaminated or otherwise hazardous environments or where very large injection volumes are used.

In most types of tests, *dilution-adjusted* concentrations of injected reactive tracers and reaction products formed in situ are computed using measured concentrations of the nonreactive tracers. Observed changes in dilution-adjusted concentrations of reactive tracers are then attributed to processes other than simple dilution. It is frequently desirable to verify that observed concentration changes are due to a targeted process and this is usually possible with proper test design. For example, the effect of sorption on reactive tracer concentrations may be detected and accounted for using nonreactive tracers that sorb similarly to targeted reactive tracers (but do not react), or by conducting tests using the reactive tracers but controlling test conditions to limit reaction progress and thus isolate the effects of sorption or other processes from those of the targeted reactions (e.g. by keeping the test solution in the formation for only a short time to limit reaction extent). The addition of various compounds that inhibit the targeted reaction can also be used to verify that observed concentration changes are in fact due to that specific reaction. For example, acetylene, oxygen, etc. are known to inhibit certain microbial processes and tests could be conducted with and without these compounds and the results compared. Alternatively catalytic compounds (such as humic acids), growth substrates, etc. may be added to increase the rate of a targeted reaction or to verify a particular reaction mechanism. There are many other possibilities as well. For example, a series of tests might be conducted to verify that reaction rates vary in a predictable way with specific system variables (tracer concentration, temperature, etc.).

Test solution composition is determined by the purpose of the test. Push-pull tests designed to determine pore fluid velocity or formation effective porosity or dispersivity are typically conducted using test solutions containing only nonreactive tracers. In these cases, other components of the test solution (pH, cations, anions, bicarbonate, etc.) may be less important and many successful tests have been conducted using test solutions prepared by simply adding salt (NaCl) to distilled or tap water. However, in other cases, one or more additional solutes may be added and these could include combinations of (1) nonreactive tracers with varying diffusion coefficients to quantify diffusion-limited mass transfer processes, (2) nonreactive tracers with varying chemical properties to quantify sorption, cation exchange, mass transfer coefficients, etc., (3) reactive tracers to quantify rates of chemical or microbiological reactions, and (4) other amendments added to control or modify test solution composition (e.g., salts to control major ion chemistry and ionic strength, dissolved gases, acids or bases to fix or buffer pH, prepared colloids, microorganisms, etc.). In many types of tests it is desirable to prepare test solutions from pore fluids collected from the test well to minimize alteration of in situ conditions. However, for other types of tests it may be more desirable to intentionally prepare test solutions with a very different composition than that of the ambient pore fluid (e.g. to poise conditions to isolate or favor a particular process) and these may be prepared in a variety of ways, depending on the test (e.g., using distilled or tap water with added salts, buffers, etc.; pore fluid from a different well, etc.). Often it is possible to collect water with the desirable composition from one well at a site and use it to prepare test solutions for use in a different well. For example, while the reactivity of hazardous materials already present in a contaminated subsurface environment might be of keen interest, it may be impractical or undesirable to add these to injected test solutions but it may be possible to use site groundwater already containing these materials. Thus, push-pull tests have been conducted with injected petroleum hydrocarbons, chlorinated solvents, radionuclides and other hazardous materials by amending site groundwater with tracers, etc. Because, at many sites, injected test solutions do not drift rapidly away from the well, it is also possible to remove most of all injected materials introduced into the subsurface during a test by extraction pumping after the test is completed. For example, mass balance calculations can be performed to demonstrate essentially complete removal of injected materials, which facilitates the use of otherwise restricted tracers. Moreover, the mass of introduced reactive tracers is typically so small that subsurface conditions are not detectably altered during a test. Thus, sequences of tests can often be usefully conducted in a single well, a set of nearby wells, or in many other combinations depending on the purposes of the study.

It cannot be over-emphasized that the composition of the test solution is the critical factor in test design as it places all results in the theoretical context needed for valid test interpretation. Of course, similar considerations are involved in the design of any laboratory experiment where it is common to add dissolved gases, pH buffers, trace nutrients, etc. and, if possible, it is often useful to review existing laboratory protocols for similar or analogous experiments during the design of a push-pull test. It should be recognized that injected test solutions will be diluted by ambient pore fluids, react with sediments, etc., and these processes will likely cause the chemical composition of the injected test solution to change during the duration of the test. In many cases, the sediment (and the "immobile" pore fluids in small pores) will dominate the biogeochemistry of the entire system because the quantities of potential reactants, catalysts, microorganisms, etc. associated with the sediment and immobile pore fluid are typically much larger than those in the injected test solution. Thus, it may not be necessary to go to extreme measures to prepare (the sometimes large volumes of) test solutions to an exact chemical recipe, since injected test solutions will immediately begin to equilibrate with the pore fluids and sediments they come into contact with during the test. For example, the pH of many subsurface environments is strongly buffered by the large quantity of reactive mineral surfaces and small differences in the pH of injected test solutions will almost always be quickly modified to reflect the pH of ambient pore fluids. But all of these factors need to be evaluated on a case-by-case basis. Also, it is always desirable to conduct a simple tracer test at the specific wells early in the test design process. This will insure that tests solutions can be injected and withdrawn at the targeted rates and can be used to determine pore fluid velocities and dilution rates. Thus, a simple preliminary tracer test in a groundwater aquifer consisting of injected tap water with a salt tracer (to create an electrical conductivity signature between injected test solution and ambient pore fluid and thus rapidly and inexpensively determine expected dilution rates) should always be used when possible before ordering tanks, reagents, sample bottles, etc.

It is usually desirable that the composition of the test solution be uniform before it is injected into the subsurface and care should be taken to insure adequate mixing prior to injection, which can sometimes be difficult given the potentially large volumes of fluids involved. For most tests it is also desirable that the composition of the injected test solution remain constant during injection and, if long injection durations are necessary (because of large test solution volumes or slow injection rates due to low permeability formations), care should be taken to avoid changes in chemical composition resulting from coagulation or precipitation of test solution components, degradation of organic tracers due to sunlight exposure, loss of dissolved components by volatilization, sorption to storage containers, exposure to the atmosphere, etc. In other cases, it may be desirable to vary the composition of the test solution in a defined way during injection (e.g. to have the concentration of a reactive solute increase during injection to examine the effects of concentration on reaction rate, etc.). In any event the composition of the test solution should be verified by on-site measurements if possible (e.g. using field kits and portable meters) prior to injection and subsequently confirmed by off-site analysis of samples of the test solution collected during injection. Care, experience, and sometimes specialized equipment are required to achieve the targeted concentrations of test solution components especially when injection volumes exceed a few hundred liters or test solution components are highly volatile, weakly soluble, or strongly sorbed. Detailed descriptions of the wide variety of methods that have been used for specific tests are contained in the examples.

Aqueous test solutions are typically prepared in large plastic or glass containers depending on the reactivity of test solution components with container materials. For example, inorganic tracers typically sorb more strongly to glass than plastic, while the reverse is often the case for organic tracers. Mixing can be achieved using mechanical stirrers, recirculating pumps, etc. although the best choice for many applications is to mix the test solution by bubbling it with compressed gases. Mixing using gases requires no electricity, can be very vigorous, and is well suited for unsupervised operation (e.g. by allowing test solutions to bubble and mix overnight prior to injection) and is particularly effective for large tanks with irregular shapes (e.g., many commercially available plastic tanks contain molded "legs" and corners that are only poorly mixed by mechanical, paddle-type mixers). Moreover, mixing with compressed gases makes it possible to modify test solution composition by varying the choice of gases, flow rates, and pressures. The choice of gases used depends on the purpose of the test but might include compressed air or O_2 to add O_2, N_2 or Ar, etc. to remove oxygen, CO_2 or N_2:CO_2 mixtures to control pH or add CO_2/HCO_3^-, He or SF_6 as nonreactive tracers, and many other gases or gas mixtures. More complicated aqueous test solutions can be readily prepared by blending separately prepared solutions during injection. For example, one solution of site groundwater with added Br^- tracer could be prepared in an open plastic tank by bubbling with N_2 gas, and a second, more concentrated aqueous solution containing a volatile reactive tracer could be prepared in a small collapsible Teflon bag. During injection the two solutions can be combined in a controlled way (e.g. using separate metering pumps) to achieve desired target concentrations of both tracers. Gaseous test solutions can be prepared by mixing gases in standard cylinders, collapsible bags, or other containers using known volumes, partial pressures, or gas flow rates to achieve the desired concentrations of all gases. Gas mixtures can be similarly introduced into aqueous test solutions by bubbling two or more gasses at flow rates proportional to the targeted concentrations. A simpler approach is to pre-mix the gases prior to bubbling. For example, gas mixtures can be easily prepared in a single compressed gas cylinder by transferring gases from single gas cylinders sequentially to achieve the desired partial pressures of each gas in the mix. Then the mixture can be bubbled through the test solution in the ordinary way to achieve the targeted dissolved concentrations of each gas (review the concepts of partial pressures and Henry's Law and consult with a local gas supplier to obtain the high pressure transfer tubing required). It is also important to consider the choice of gas on other components of the test solution. For example, if bubbling a dissolved gas is used to add a reactive tracer to the test solution (e.g. the addition of acetylene to inhibit denitrification) it should be recognized that the concentrations of other dissolved gasses will decrease, which may result in unintended changes in test solution composition. For example the unintended removal of dissolved CO_2 from a test solution by bubbling with any other gas may result in decreased pH, which could result in precipitation formation or many other chemical changes. Some test solution components may be photoreactive and thus it may be necessary to protect the test solution from light prior to injection by covering the mixing tank with a suitable material.

The volume of injected test solution should be measured and samples of the test solution should be collected periodically so that the actual composition of the injected test solution is known. Liquid volumes can usually be measured visually as most commercial tanks have volume markers molded into tank walls or by periodically measuring pumping rates and elapsed times. Gas volumes may be measured using ideal gas law calculations based on measured tank pressures before and after injection, or using in-line gas flow meters to measure gas flow rates and elapsed times. Liquid solution injection can be accomplished in a variety of ways including (1) using pumps to transfer the test solution to the well, or (2) draining the contents of the supply tank into the well by gravity or using a siphon. Gas injections are usually conducted using pressure in the compressed gas cylinders. If necessary, care should be taken to avoid changing the composition of the test solution during injection (e.g., by avoiding contact with oxygen in the atmosphere or by unintentional variation in pumping rates). For some types of tests, the injection rate is required for data analysis and this can be monitored volumetrically or with suitable flow meters. The injection rate is ultimately limited by the formation permeability and the length, diameter, etc. of the borehole, well screen, etc. Care should be taken to avoid excessive buildup of fluid pressure in the formation during injection as this will generally result in higher dilution losses than if injection rates are smaller (high pressures can also fracture formations, damage wells, overflow casings, etc.). For example, in unconfined aquifers a substantial increase in water table elevation will deliver a portion of the test solution to the overlying vadose zone, where it will be partially retained by capillary forces and thus unavailable to subsequent groundwater sampling. Injected gases can also escape from the system (e.g. to the atmosphere) if injection pressures are high. Leaks are more difficult to detect when injecting gaseous test solutions but field procedures can be easily tested by conducting a short-duration tracer test and performing mass balance calculations.

Test solutions can be injected into the formation through the walls of an open borehole or through the screen of a well or drive point. If desirable, straddle packers may be used to interrogate specific depth intervals within the well or borehole; multi-level wells or clusters of wells completed at different depths can also be used for this purpose. Recording the total volume of test solution injected during the test is desirable for use in mass balance calculations and this can usually be determined volumetrically using a calibrated tank (most commercially purchased tanks have volume marks on the side and these can simply be recorded as a function of time). It should be noted that if injection rates and pore fluid pressures are recorded during a push-pull test, formation conductivity can be determined with no additional effort using conventional methods of pumping test data analysis. Thus, in addition to measuring water levels on a tank during injection, it may be desirable to measure water levels or gas pressures as a function of time during injection and/or extraction pumping either manually using a water level indicator or in-line pressure gage or automatically using a pressure transducer and datalogger so that data may be interpreted to obtain an estimate of conductivity. Down-hole instruments can also be used to record other variables such as temperature, electrical conductivity, dissolved oxygen, pH, etc. during a test if desirable (e.g. if electrical conductivity or ion specific electrode measurements are being used to measure changing concentrations of a nonreactive tracer). However, critical review of test plans should be performed to avoid making field procedures overly complex unless absolutely necessary for test interpretation. Thus, it may be desirable to conduct more tests with less rigorous experimental controls and with less collection of ancillary data than a few complex tests if the goal is to assess system behavior over a large field site. Many subsurface environments are highly heterogeneous with properties varying widely over short distances and may be more effectively characterized by logistically simple tests conducted in many locations than by a few complex tests conducted in only a few locations.

A variety of approaches may be used to monitor the composition of the test solution/pore fluid mixture within the formation after injection is completed and the choice of approach depends on the objective of the test. It is important to recognize that samples collected after injection represent a continuously changing mixture of injected test solution and ambient pore fluid, with the proportion of injected test solution in samples typically decreasing (as the proportion of ambient pore fluids increases) over time. For some types of tests, it is desirable to extract the test solution/pore fluid mixture at a constant rate and this can be accomplished using pumps or bailers. Mass balance information can be obtained if the extracted volumes associated with each sample are recorded (e.g. by pumping a defined volume prior to the collection of each sample). As in pumping tests, extracted water should not be disposed at the surface near the test well to avoid the potential for pumped fluids from re-entering the formation near the test location and possibly confounding test results. For tests aimed at determining reaction rates, continuous extraction pumping may not be desirable and instead, the composition of the test solution/pore fluid mixture can be monitored by periodic discrete sampling events spaced hours, days, weeks, or even months apart as needed to monitor anticipated changes in test solution composition. In general, acceptable results can be obtained using conventional sampling protocols (i.e., some amount of purging followed by sample collection) but it is best that consistent procedures are used. For example if a specified volume of purge water is discarded prior to sample collection, that procedure should be followed for all sampling events. Similarly, pumping or bailing rates should be the same for all sample events to insure that the same portion of the near-well environment is sampled each time.

Obviously, the specific details of the sampling protocol should match the requirements of the intended chemical or microbial analyses (i.e., the required sample volumes, types of container, use of preservatives, handling and storage requirements, etc.). If injected test solution volumes are very small (on the order of the combined volume of all samples collected), it may be necessary to carefully record times, volumes, etc. for all samples recorded but in most cases the combined volume of all samples is negligible compared to the volume of injected test solution. It is sometimes difficult to predict when samples should be collected because the rates of various processes (i.e. dilution rates, reaction rates, etc.) are unknown (which is why the test is being conducted in the first place!). In these situations it is desirable to *over-sample*, to collect samples at a much higher frequency than necessary to insure that major changes in tracer concentration are captured during the test. Of course, analyzing such a large number of samples may be prohibitively expensive and for this reason an efficient approach is to analyze nonreactive tracer concentrations on all samples first and then select a subset of the remaining samples for the typically more expensive reactive tracer analyses using the nonreactive tracer data as a guide. Using this approach, useful dilution-adjusted concentration profiles can be obtained with as few as 10 samples analyzed out of perhaps 100 collected. On-site measurements of nonreactive tracer concentrations can also be used to guide the collection of more-expensive samples for reactive tracer analyses. Thus, electrical conductivity, pH, Cl^- or Br^- concentrations by ion specific electrodes and meters, and similar quasi-quantitative field methods can be of great use in identifying times for collecting more expensive or labor intensive samples. Although a number of field test kits have been developed for analyzing e.g., nitrate, sulfate, some metals, etc. by spectrometry or similar technique it is always best to collect a complete suite of samples for laboratory analysis to insure that the highest quality quantitative determinations are performed. It is also often prudent to collect duplicate samples to preclude accidental sample loss during shipping, storage, etc.

An important feature of the push-pull test approach is that it is usually possible and desirable to conduct tests in multiple wells simultaneously so that spatial heterogeneity in system behavior can be quantified or to meet other test objectives (i.e. tests conducted in a particular sequence or within a specified time frame, or to reduce overall testing cost). This can often be done with little additional effort or cost compared to performing a single test. If multiple tests are performed simultaneously, it is also possible to include replicate determinations and various controls or alternate treatments into the overall test design. Simultaneous testing is possible because of the modest field equipment required to conduct most types of push-pull tests and because the zone of influence of individual tests is typically quite small (e.g. injected test solutions may travel only a meter or two from the test well) so that tests conducted in nearby wells rarely interfere with each other. With sufficient containers for test solution preparation, injection equipment, chemicals, and sample bottles it is possible for an experienced field person to initiate several tests in a single day. For example, at one field site, rates of multiple microbial processes (denitrification, sulfate reduction, iron reduction, and metal reduction) were quantified using test solutions prepared from site groundwater that had been pumped from each test well and stored in plastic 200 L drums placed next to each well. After the tracers were added, test solutions in all drums were mixed overnight by bubbling with compressed gases and then injected using simple plastic tubing as siphons to deliver the test solution from each drum to its adjacent well. In this example, 10 tests were initiated by a single person in 2 days (the wells were then sampled periodically over the next 10 weeks using conventional sampling techniques).

Several methods for push-pull test data analysis have been developed and these are described in detail in the literature cited in the examples. Generally these methods begin by plotting *concentration profiles* (i.e. relative concentrations vs. time) for each tracer, where the *relative concentration* is defined as the measured tracer concentration in a sample divided by the concentration of the same tracer in the injected test solution or ambient porefluid. To compute reaction rates and certain other biogeochemical parameters it is usually necessary to plot *dilution-adjusted concentration profiles* (i.e. the measured concentration of a reactive tracer divided by the relative concentration of the nonreactive tracer) versus time. Generally, formation properties, reaction rates, etc. are typically determined by fitting the concentration profiles with a specific model, depending on the type of test, as shown in the examples. Most calculations utilize empirical or analytical solutions to the flow and transport equations and can be performed in a spreadsheet program but numerical flow and transport codes and geochemical modeling software are also sometimes used. It is interesting to note that the unique format of the push-pull test (i.e. injection followed by extraction at the same location) typically produces "well-behaved" concentration profiles (i.e. single peaks, smooth, and monotonic data distributions with time) that are easy to interpret and fit using simple models. This is so because flow reversal tends to decrease the effects of system heterogeneities on fluid flow and tracer transport. This behavior also affects the overall test costs because useful concentration profiles can almost always be obtained with relatively few extraction phase sampling events.

Chapter 2
Methods

2.1 Injection/Extraction Possibilities

Push-pull tests can be conducted using any facility or device that makes it possible to inject and extract pore fluids from the formation (Fig. 1.2). Thus, tests may be conducted in open boreholes, screened intervals of conventional monitoring wells, sampling ports of multi-level monitoring wells, drive-points wells, piezometers, or through drive points inserted into the sidewalls of an open excavation. Tests may be conducted above or below the water table, at any depth, and in any type of geologic formation. Tests may be conducted in terrestrial subsurface environments or in saturated sediments that lie beneath lakes, rivers, estuaries, the sea floor, etc. Push-pull tests are most suitable for tests conducted in porous media where flow is laminar but push-pull tests have also been conducted in deep lakes and other surface water bodies where weak turbulent mixing limits dilution losses of injected test solutions. Small-scale push-pull tests have been successfully conducted using the simplest equipment, such as plastic syringes injecting test solutions through "well screens" formed of syringe needles manually inserted into saturated or submerged sediments.

Test locations are typically selected to meet test objectives but it is common and usually desirable to use existing wells, etc. as this reduces the costs associated with well installation, pre-test monitoring, etc.

2.2 Volume of Injected Test Solution

The volume of injected test solution is almost always selected to be several times larger than the computed 'dead volume' of the injection well to insure that most of the injected test solution penetrates into the formation, rather than simply residing in the well or borehole. The dead volume is calculated from the well or borehole diameter and depth, the length of screen/casing, pump and tubing volume, etc. and

J.D. Istok, *Push-Pull Tests for Site Characterization*, Lecture Notes in Earth System Sciences 144, DOI 10.1007/978-3-642-13920-8_2,

should be increased to reflect the influence of filter/sand packs, formation disturbance due to drilling, etc. as appropriate. The 'dead volume' is thus an important factor controlling test duration and cost. Clearly, much larger test solution volumes must be prepared for tests in deep, large-diameter wells, which typically contain larger dead volumes than shallow, small-diameter wells. Larger test solution volumes increases test cost by increasing the size of tank(s), quantities of reagents, and the times required to collect, mix, and inject test solutions. Fortunately, 2 in. and smaller diameter wells are common at most sites and small diameter wells are more commonly being installed to minimize purge volume requirements for routine sampling. However, 4 in. diameter and larger wells are still common, especially at sites where they were installed for active pumping (pump-and-treat remediation, municipal or domestic water production, etc.). Also, at some sites computed volumes for even small diameter wells can be very large if depths are greater than ~100 m. When computed well dead volumes are deemed excessive they may sometimes be reduced through the use of 'straddle packers' or similar systems that isolate a portion of the well screen for testing (Fig. 1.2). These are installed in the well to the desired depth and then inflated (or otherwise activated) from the surface using e.g., compressed air. When inflated, the packers prevent vertical movement of water across the packers within the casing, thus reducing the dead volume to the volume of water within the packed interval and the tubing necessary to carry fluids to/from the surface. Straddle packers may also be used as one option when the goal is to conduct a test at a specific depth within a formation. It should be noted that although straddle packers isolate the portion of the well screen through which fluids enter and exit the well, flow to/from a packed interval is not strictly horizontal, especially if sand packs or disturbed zones created by well installation increase the vertical permeability of the formation near the well. Other options for conducting push-pull tests at specific depths include the use of wells/drivepoints, etc. installed at different depths, and the use of dedicated multi-level monitoring wells (Fig. 1.2).

The main factor controlling selection of the test solution injection volume is the desired formation volume to be interrogated during the test. This 'representative elementary volume' (REV) should be selected to insure that test results are relevant for the purposes of the investigation. The selection of a suitable REV is ultimately subjective but in general incorporates all relevant site knowledge regarding geology, formation properties, fluid flow patterns, geochemistry, microbiology, etc. As a rough guide, an REV of ~1 m^3 is usually sufficient to capture most important biogeochemical processes occurring in the vicinity of a well. However, much larger and much smaller REVs have been selected in previous studies (see papers cited in Chapter 3) and the selection is necessarily site and problem specific.

The total volume of the interrogated zone (solids plus pores) can be estimated using:

$$\text{Sample volume} = \frac{\text{Injected volume of test solution} - \text{``dead volume''}}{\text{Formation effective porosity}} \tag{2.1}$$

Equation 2.1 can be modified to account for the volume of annular space between the well casing and borehole wall, the presence of sand packs/filters in the annular space, etc. It should be recognized that the exact three-dimensional shape of the interrogated zone is never known due to heterogeneity in the formation's porosity, permeability, etc., which causes the test solution to penetrate into the formation irregularly, varying with depth and distance from the well. The shape of the interrogated zone will also be strongly influenced by layering, the presence of preferential flow paths (e.g., fractures), density differences between injected test solutions and ambient pore fluids, and by the amounts and kinds of disturbances created by borehole construction and well installation. In practice, the dead volume is first calculated for the test well and then the injected volume of test solution is adjusted to obtain the desired formation volume (subject to the uncertainty in effective porosity). Except for the minimum volume imposed by the dead volume, the injected test solution volume can be increased to achieve any desirable sample volume. Tests have been conducted to interrogate sample volumes of up to ~100,000 m^3. However, in most cases, acceptable results can be achieved with sample volumes of 200–1,000 L, which require injection volumes of ~50–500 L, depending on the effective porosity of the formation. Logistical considerations also usually play a large role in selected test solution volumes (Fig. 2.1). The size of available tanks to store and mix test solutions, the weight of tanks and water, the ability to collect sufficient groundwater to fill the tanks, the costs of required chemicals such as tracers, buffers, and dissolved gases, regulatory limits on injection volumes, time required to inject large test solution volumes, and storage or disposal of extracted water during sampling or pumping, are a few of the factors that typically restrict test solution injection volumes to a few 100 L.

A rough estimate of the distance from the well that the test solution penetrates can be computed from the injected test solution volume by assuming that the penetrated portion of the aquifer is cylindrical so that:

$$r_{max} = \sqrt{\frac{V}{\pi h}} \tag{2.2}$$

where V is the sample volume computed from Eq. 2.1, and h is the saturated thickness.

This equation is somewhat useful in providing a rough estimate for the relative volumes of test solution required to penetrated specified distances from the test well but cannot provide an accurate estimate because it ignores the effects on fluid flow of dispersion, diffusion, heterogeneity in formation hydraulic properties, fluctuating water levels, which change saturated thickness, etc. One way to increase the penetration distance without increasing the volume of injected test solution is to reduce the "dead volume" e.g., by injecting test solutions through only a portion of the well screen or borehole wall using a pair of straddle packers as discussed above. This approach is particularly desirable in large-diameter, deep wells. Equations 2.1 and 2.2 clearly show the logistical advantages of using small diameter wells for

Fig. 2.1 Field push-pull test photographs showing various methods used to prepare and inject test solutions. Clockwise from *upper left*: separately prepared tests solutions in 50 L plastic carboys being combined and injected into a well using a peristaltic pump; 500 L test solution prepared in plastic tank; 240,000 L test solution being transferred from tanker truck to large plastic tank prior to injection; test solution in 200 L plastic drum being injected into monitoring well using a siphon; 50 L test solution in plastic carboy being injected into monitoring well by gravity drainage

push-pull testing; smaller diameter wells have smaller "dead volumes" (all other factors being equal) and larger sample volumes and penetration distances can be obtained in a small diameter well using a given test solution injection volume compared to a large diameter well.

Another factor in selecting the injection volume is the anticipated dilution rate. After injection, the test solution will be diluted with ambient pore fluids as the test solution flows away from the injection location. Concentrations of injected tracers will decrease and eventually reach nondetectable or background levels. The ability of a push-pull test to provide useful information is determined to a large degree by the relationships between injected test solution volume, the dilution rate, and the time required for the test solution to remain in the formation to provide useful results. For example, if the goal is to measure the rate of transformation of an injected reactive tracer, it will be necessary to keep the injected test solution in the vicinity of the well long enough to allow the reaction progress to be observed. This is most easily achieved if reaction rates are large and dilution rates are small. In that case, the injected test solution volume can be selected solely to meet other scientific or logistic objectives. However, if reaction rates are anticipated to be small and/or dilution rates are anticipated to be large, it is possible that the injected test solution will migrate away from the vicinity of the well too quickly to provide any useful information. In this case, the injection volume can be increased beyond that determined by other factors in an attempt to compensate for the higher anticipated dilution rate. Simple calculations with Darcy's Law, effective porosity, and assuming a cylindrical zone of influence are usually adequate to estimate a suitable injection volume. Of course these calculations are only approximate as the dilution and reaction rates are not known a priori (otherwise there would be no point in conducting the test!). However, even approximate calculations based on crude rate estimates are usually sufficient to select a test solution volume that insures a successful test. Nevertheless, it is usually desirable to conduct a simple test with a nonreactive tracer to estimate the dilution rate prior to injecting expensive or difficult-to-prepare test solutions or performing time-consuming or expensive sampling and analyses. In many cases, simple injections of tap water followed by periodic measurements of electrical conductivity or some similar field parameter are sufficient to determine how long injected test solutions will reside in the formation near the test well. For example, if tap water is injected, the changing chemical composition of the test solution may be indicated by a changing electrical conductivity that will vary between the injected tap water and ambient groundwater at a rate that is roughly proportional to the dilution rate. Of course for this type of test to succeed requires a diagnostic chemical signature between injected and ambient porefluids.

The desired test solution volume is also usually modified to accommodate various site-specific logistical factors. For example, containers available for preparing test solutions may have a fixed size (e.g., 200 L drums are common at many contaminated sites and farm supply stores sell plastic tanks in incremental sizes up to ~5,000 L) or makeup water may have to be hauled in trucks with fixed weight carrying capacity. Costs of tracers, compressed gases, etc. also increase as the test solution volume increases and these costs must also be included in the decision making process. Another important cost factor is the time required to inject the test solution, which is ultimately limited by the formation permeability and by the diameter and length of the screened interval. While injection rates of ~1 L/min are common, the possible range is extremely wide, from a few mL/min to tens of L/min. Obviously, injecting large test solution volumes in small diameter wells installed in low permeability materials may require many days and can be the rate-limiting factor in determining overall test duration and cost. Moreover, push-pull tests used to determine pore water velocities, mass transfer coefficients and sorption characteristics may require the extraction of between two and five times the volume of injected test solutions and this can add considerably to the time required to complete these types of tests. However, with experience, a balance of all of these factors can be readily achieved to meet test objectives.

2.3 Test Solution Composition

The type of test is used to determine the composition of the injected test solution. The first decision is to select a suitable liquid or gas to which various chemical amendments will be added. In saturated zone tests, test solutions are typically prepared from site groundwater in an attempt to insure that test results represent in situ conditions as closely as possible. Using site groundwater preserves the concentrations of major and minor ions, dissolved gases and other volatile constituents that may play a role in the specific chemical or biological reactions or processes under investigation. For example, site groundwater could be extracted from each test well, modified to the minimum extent possible by adding only necessary tracers, etc. and then injected into the same well. The knowledge and experience of the investigator is used to select those features of the site groundwater that should be preserved during test solution preparation and these will influence the details of above-ground operations. For example, test solutions may be collected in inflatable bladders rather than tanks to avoid the loss of volatile components and exposure to atmospheric oxygen or in insulated containers to avoid temperature changes while test solutions are manipulated above-ground. Of course all of these provisions will create logistical difficulties and increase the time and cost of testing. It should be mentioned that the generally high reactivity of subsurface sediments will quickly modify some attributes (e.g., pH) of injected test solutions and that the contribution of trace reactive components in site groundwater are typically overshadowed by the much higher concentrations of reactive species associated with mineral surfaces, immobile water, etc. present in the aquifer. For this reason, the need to take extreme care to preserve all features of site groundwater during test solution preparation should be critically evaluated. As a simple example, many contaminated aquifers may contain large quantities of reduced iron (mmol to mol Fe^{2+}/kg of sediment) produced by microbial iron reduction. This Fe^{2+} will rapidly react with any trace dissolved O_2 remaining in injected test solutions, removing O_2 from solution:

$$Fe^{2+} + H^{+} + \frac{1}{4}O_2 \leftrightarrow Fe^{3+} + \frac{1}{2}H_2O$$

Similar reactions may result in the precipitation of other reduced metals, sulfides, etc. Thus, in anaerobic environments taking extreme care in avoiding the introduction of trace amounts of O_2 (~μmol/L) during above-ground test solution preparation (which can be very difficult) may not be warranted. Also, most chemical reactions of interest are generally insensitive to changing concentrations in major inorganic ions, etc. so that attempting to create a synthetic groundwater with the exact composition of ambient groundwater may not be worth the cost or effort.

In many cases, test solutions prepared from distilled water, tap water, or synthetic groundwater are used in place of site groundwater (typically to avoid some of the logistical issues identified above or for regulatory reasons) where it is believed that test results will be largely unaffected by the exact chemical composition of the injected test solution. Some examples are nonreactive tracer tests performed to characterized ambient pore fluid velocities, effective porosity, dispersivity, and mass transfer coefficients. Using these types of makeup waters can greatly reduce the time and cost of field testing; test solutions prepared from tap water or other readily available bulk water sources have been widely and successfully used for many types of tests. Similar considerations pertain when selecting gases to prepare test solutions for vadose zone tests; the trace impurities present in lower quality and lower cost readily available compressed gases typically have no effect on test results compared to using higher-purity, special order gases.

The number, type, and concentration of tracers included in the injected test solution depend on the purpose of the test. Tests to determine pore fluid velocity, formation dispersivity, mass transfer coefficients, etc. are typically conducted using only nonreactive tracers, while tests to determine reaction rates are conducted with one or more additional reactive tracers as needed to interrogate the targeted process.

Many nonreactive tracers are available and include inorganic anions (Br^-, Cl^-, etc.), organic anions (chlorinated or fluorinated benzoic acids), dissolved gases (SF_6, He, Ar, Ne), and many other possibilities. Selecting a nonreactive tracer depends on a number of factors including the chemical composition of ambient pore fluids, the availability and cost of suitable analytical methods, tracer costs, and the objectives of the test. The nonreactive tracer should not be present at high concentrations in the ambient pore fluid as this will complicate the calculation of dilution-adjusted concentrations. However, if nonreactive tracer concentrations in ambient groundwater is high, it may be possible to inject test solutions with no nonreactive tracers and still calculate dilution-adjusted concentrations (e.g., tests conducted in salty groundwater, seawater, or brines). Thus, Cl^- and Br^- are generally suitable choices for nonreactive tracers for use in tests conducted in low ionic strength groundwaters, which typically have low Cl^- and Br^- concentrations, but may not be suitable for tests conducted in groundwater containing higher concentration of these ions or perhaps other ions that create analytical interferences. For example, the use of Cl^- as a nonreactive tracer is problematic where the subsurface is impacted by seawater or brine (e.g., in coastal locations or certain depositional environments), which typically have large Cl^- concentrations. Similarly the presence of high Cl^- concentrations in ambient porefluids may complicate accurate quantification of Br^- added to injected test solutions due to analytical interferences (e.g., by ion chromatography). It is always desirable to verify analytical methodology in the laboratory (e.g., by spiking nonreactive tracer into a sample of ambient porefluid) prior to field testing. Inorganic tracers are typically analyzed by ion specific electrodes, ion chromatography, or inductively-coupled plasma mass spectrometery. Organic tracers are typically analyzed by flourometry, spectroscopy, ion chromatography, gas chromatography, or liquid chromatography, while gas tracers are typically analyzed by gas chromatography.

The concentration of nonreactive tracer(s) used should be given careful consideration. As a preliminary guide, many tests utilize injected tracer concentrations that are approximately 100 times larger than the analytical detection limit. For example, if Br^- analysis by ion chromatography has a detection limit of ~1 mg/L then a suitable Br^- concentration for the injected test solution would be ~100 mg/L. This allows for the accurate determination of tracer concentration as injected test solutions are diluted by ambient pore fluids during the test. When the analytical detection limit is reached, samples collected from the well will consist of ~1 % injected test solution and ~99 % site groundwater. When dilution rates are anticipated to be large (i.e. ambient porefluid velocity is anticipated to be large) it may be tempting to increase tracer concentration and some tests have been conducted with very high tracer concentrations (up to 250,000 mg/L!). However, experience has shown that the use of such high tracer concentrations is almost always counter-productive and results in *increased* dilution losses due to buoyancy induced vertical sinking of injected test solutions caused by density differences between the test solution and the ambient pore fluid. It also may be necessary to consider the effects of density differences on the degree of mixing (or lack of mixing) that occurs between injected test solutions and pore fluids within the well casing and screen. The overall result is larger apparent dilution losses of injected test solution. High concentration and therefore high density test solutions will also sink to the bottom of wells and boreholes and cause injected test solutions to preferentially penetrate deeper portions of the formation. The latter effect can be partially ameliorated by mixing the contents of the well casing during test solution injection but density driven flow will still occur in the formation whenever there is a significant density difference between injected test solutions and ambient pore fluids.

It should also be noted that the addition of high concentrations of nonreactive tracers can sometimes have additional unintended consequences. For example, the addition of high concentrations of Na^+, Ca^{2+}, Mg^{2+}, etc. (present in salts used as a source of Cl^- , Br^- , etc.) to the subsurface can alter the charge balance on clays and other mineral surfaces resulting in decreased porosity and permeability of the formation or in the release of ions from mineral surfaces (e.g., heavy metals, radionuclides, some organics) by cation exchange and/or mass action. High salt concentrations may also decrease the solubility of other test solution components in sometimes complex and unpredictable ways, making it difficult to maintain the desired overall composition. For these reasons, the consequences of using test solution tracer concentrations greater than a few hundred mg/L should be critically evaluated and rigorously tested in the laboratory prior to field implementation.

The choice of reactive tracer(s) is also a critically important part of push-pull test design and these are determined by the purpose of the test. As the Chapter 3 and Examples sections describe, tests have been conducted to examine the reactivity of a wide variety of such tracers including dissolved gases (e.g. oxygen and methane), organic contaminants (e.g. chlorinated solvents such as trichloroethene and petroleum hydrocarbons such as toluene), inorganic contaminants (e.g. heavy metals such as chromium and radionuclides such as uranium and technetium), and many others. In addition to the considerations listed above for nonreactive tracers, selecting the type and concentration of reactive tracers may require a number of additional factors. For example, the reactivity of many of these tracers will depend on concentration and the presence of other components in solution. For example, the reductive dechlorination of a chlorinated solvent typically depends on the concentration of both the electron acceptor and electron donor (e.g. trichloroethene and hydrogen) as well as other factors and these must be considered in designing test solutions aimed at quantifying rates of reductive dechlorination. Similarly, H^+ and HCO_3^- participate in many chemical and biological reactions and modifying or maintaining ambient concentrations of these ions may be desirable. The Examples should give preliminary guidance and it is also desirable to review protocols used in laboratory experiments to identify the specific combination of nonreactive tracers and other test solution components required to interrogate a specific reaction.

2.4 Test Solution Preparation

Test solution preparation is largely a matter of collecting a sufficient quantity of makeup fluid and modifying it in various ways to obtain the desired composition before it is injected into the subsurface. For aqueous test solutions, this usually involves first filling a glass or plastic container with makeup water and then mixing in various chemical or other amendments. It is important to classify the desired test solution components by their physical form (solid, liquid, gas) and chemical properties (solubility, volatility, etc.) in order to select an appropriate container and method of mixing. For example, anionic tracers like Br^- and Cl^- are added as salts (e.g., NaCl, KBr) and these are highly water soluble, not expected to sorb to glass or plastic, and are nonvolatile. Thus, adding these tracers to aqueous test solutions involves simply adding the desired amount of the appropriate salt (NaBr, NaCl, etc.) to a known volume of water in any suitable container and mixing by any convenient method until a constant composition is obtained. However, even in this simple case, a number of additional considerations may arise. For example, if test solutions are prepared from site groundwater it should be recognized that the addition of salts and the agitation of mixing may cause undesirable physical/ chemical affects. For example, the addition of salts may decrease the solubility of other aqueous species and may lead to the precipitation of various mineral phases. Also exposure of the makeup water to the atmosphere and the agitation required for mixing may increase the dissolved oxygen content of the test solution and lead to the oxidation of reduced species (e.g. HS^-, Fe^{2+}) and the precipitation of oxides, sulfides, etc.. Mixing may also lead to the loss of volatile components and these also may have undesired consequences. For example, loss of dissolved CO_2 may lead to an increase in pH, decrease in HCO_3^-, decreased solubility of various metals, precipitation of various mineral phases, etc. Temperature and pressure changes during test solution preparation may sometimes have similar undesirable or unanticipated effects (e.g., degassing caused by depressurization when ambient groundwater is brought to the surface). The combined effects of all of these processes are difficult to predict with complete confidence. Nevertheless, in the vast majority of tests it is a relatively simple matter to keep the magnitude of these effects small and if a particular effect is deemed undesirable for a particular type of test, field

Fig. 2.2 Example simple setup for mixing test solutions in plastic tank using compressed gas supplied by the cylinder to porous stones placed in the bottom of the tank containing the test solution. After mixing, the test solution is injected into the well using the peristaltic pump

procedures may be easily modified to eliminate that effect and produce a test solution with the desired composition as discussed below.

The dissolved gas composition of an aqueous test solution can be controlled in several ways. For example, if the goal is to maintain the dissolved gas composition of ambient groundwater during test solution collection and preparation, site groundwater can be collected in inflatable, gas-impermeable bags or bladders to prevent the loss of volatile groundwater components or the introduction of atmospheric gases. If instead the goal is to change the dissolved gas composition in a prescribed manner, this can often be accomplished by bubbling compressed gas mixtures through the makeup water. One effective way for doing this is place several porous stones (e.g., of the type used in home aquariums) in the bottom of the container and connecting these to compressed gas cylinders or other sources of compressed gases by plastic tubing (Fig. 2.2). For example, a small pump or compressor can be used to deliver air to the stones and bubble air through the makeup water, while compressed gas cylinders can be used to deliver O_2, CO_2, SF_6, H_2, etc. Gas bubbling has proven to be an effective method for mixing test solution components and the number of stones and the gas pressure can be easily increased to mix test solutions of any size. Compressed gas is often more convenient than mechanical mixers because a single standard compressed gas cylinder can be used to simultaneously mix many large tanks of water by simply extending/splitting gas delivery lines as needed. Compressed gas mixing requires no electrical power and many commercial vendors for a wide variety of compressed gases are available so that obtaining gas cylinders is usually not difficult, even at remote sites. Using compressed gas allows test solutions to be mixed unattended for many hours or days if necessary, which can substantially increase overall test productivity. A common approach is to mix solutions overnight prior to the start of test solution injection and to continue bubbling during injection.

Perhaps most importantly, using compressed gases to mix aqueous test solutions makes it possible to control certain aspects of the chemical composition of the test solution. This is done by specifying the flow rate(s) or partial pressure(s) of the supply gases to achieve the desired dissolved gas concentrations. For example, bubbling with compressed air or oxygen can be used to add dissolved oxygen to test solutions; similarly bubbling with compressed N_2 or Ar can be used to remove

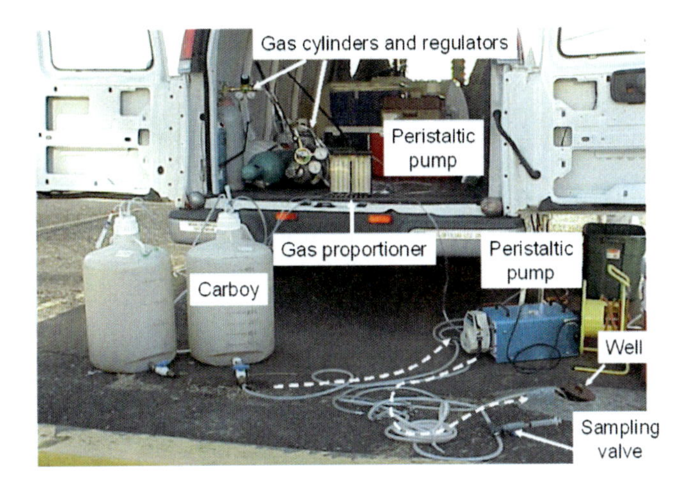

Fig. 2.3 Test solution prepared by bubbling three separate gases simultaneously. Gas flow rates from the cylinders to the porous stones in the carboy are controlled using the gas proportioner, which contains a separate float-type flowmeter and adjustable needle valve for each supply gas. Dissolved gas concentrations are computed from gas properties, flow rates, and temperature. After mixing, the test solution is injected into the well using the peristaltic pump; samples are collected to verify concentrations of all components

dissolved oxygen from test solutions. Specified dissolved oxygen concentrations between zero and saturation can be easily achieved by bubbling with a prepared gas mixture with given partial pressures of O_2 or by bubbling two separate gas streams (O_2 and another gas, e.g., N_2) at proportional flowrates through the test solution (Fig. 2.3). Specified concentrations of other dissolved gases (He, Ne, SF_6, H_2, CH_4, C_2H_2, CO_2, etc.) can also be achieved by bubbling mixtures of those gases through the test solution. Estimates for the conditions needed to achieve the desired dissolved gas composition can be obtained from tables of gas solubility, Henry's Law constants, etc. for a given partial pressure and temperature; but these estimates should be confirmed prior to field deployment. It should be mentioned that unusual gas mixtures may be easily and inexpensively prepared by first combining various partial pressures of pure gases in a suitable empty cylinder using a transfer device built from high pressure tubing and appropriate regulators. Once the gas cylinder containing the mixture is prepared, the gas mixture from that cylinder is bubbled through the test solution as usual to achieve the targeted concentrations of all components.

It is important to recognize that bubbling dissolved gases through the test solution may result in other (perhaps unintended) chemical composition changes. For example, if test solutions are prepared from site groundwater, the concentration of volatile contaminants and dissolved gases will be reduced (unless these are present in the compressed gases bubbled through the test solution). This may be desirable in some situations, (e.g., to remove dissolved oxygen), but not in others (e.g., if the reactivity of volatile components of ambient groundwater will be studied during the test). It is important to note that loss of dissolved CO_2 may change the pH and other characteristics of the test solution (e.g., pH affects the solubility of many dissolved metals). For this reason, it is often desirable to include CO_2 in the gas mixture (e.g., an anoxic test solution might be prepared by bubbling a gas mixture of 80 % N_2 and 20 % CO_2) and to monitor pH during mixing and injection.

Adding the various chemical amendments followed by mixing with compressed gases is a convenient method for preparing aqueous test solutions in many cases. However, this approach may not be suitable if some components of the test solution are toxic, expensive, or can only be purchased in liquid form. For example, while it may be desirable to prepare aqueous test solutions containing known concentrations of a volatile organic solvent it may not be practical to bubble solvent vapors through the test solution because of costs and difficulties in preparing and pressuring a suitable gas phase, capturing and treating potentially toxic gas streams leaving the tank, and achieving a constant composition. Also, to achieve certain chemical compositions it may be necessary to bubble potentially flammable or explosive gas mixtures through the aqueous test solution (hydrogen, acetylene, oxygen, methane). Moreover, some specialty gases are very expensive and it may not be cost effective to bubble sufficient quantities of gas through the test solution to achieve targeted concentrations.

In these cases, alternate approaches may be used to prepare aqueous test solutions with specified concentrations of dissolved gas or other volatile components. One approach is to use a coil of gas permeable tubing immersed in the test solution. The tubing is connected to a compressed gas cylinder. When pressurized, gas is transferred to the test solution by diffusion through the tubing walls without bubbling. A related approach is to add prepared solid phases (e.g., activated carbon) or nonaqueous phase liquids (e.g., mineral oil) containing the targeted compound directly to the makeup water. In this case, gas is transferred by diffusion from the interior of the solid or nonaqueous phase liquid to the makeup water. These approaches can be effective and extremely convenient but they usually require preliminary laboratory testing to insure that desired uniform concentrations are achieved (e.g., diffusion is a rather slow process and diffusion rates will depend on temperature, material properties, and a number of geometrical factors; in any case, some form of mechanical mixing may be required to achieve uniform concentrations within a large volume).

An alternate approach for formulating aqueous test solutions with volatile components that has many practical advantages is to prepare small volumes (1–10 L) of concentrated aqueous solutions of these materials in collapsible bags (Fig. 2.4). These bags are commercially available, are constructed of various inert and gas-impermeable materials (e.g., Teflon), and are available in a wide variety of sizes. The bag is first filled completely with water leaving no headspace. Then a syringe is used to inject known quantities of gases or other substances into the bag, which can then be mixed by hand (e.g., by "gentle scrunching"). For example, an aqueous test solution containing a known concentration of dissolved ethene could be prepared by injecting the required volume of ethene gas (at a specified pressure and temperature) into a collapsible bag filled with site groundwater using a gas-tight syringe. Additional test solution components can be sequentially added as needed to

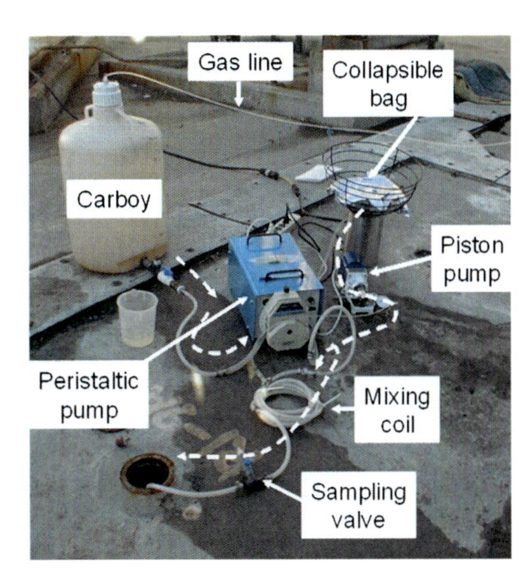

Fig. 2.4 Nonvolatile test solution components are prepared in the plastic carboy; volatile test solution components are prepared in the collapsible Teflon bag. The two solutions are combined during injection to obtain the desired overall composition. The "mixing coil" is used to insure complete blending of the two solutions before they are injected. Test solution composition is verified by analyzing samples collected from the sampling valve

achieve the desired overall test solution composition. As the test solution is pumped from the bag during injection, it collapses, maintaining a zero head space condition within the bag and thus preventing losses of volatile test solution components. Other advantages of this approach are that (1) only small quantities of gases or other solutes are required, and (2) the entire system is closed, minimizing losses of volatiles to the atmosphere, chemical exposure of field personnel, etc.

It is sometimes desirable to combine various approaches for preparing aqueous test solutions. For example, nonvolatile test solution components can be prepared in a tank open to the atmosphere and mixed with compressed air or other gases, while volatile test solution components can be prepared as aqueous solutions in small collapsible bags as described above. The two solutions, prepared separately, can be combined (e.g., using two pumps or a single pump with two pump heads) during injection to achieve a test solution with the desired overall composition (Fig. 2.4). By calibrating the pumps at the correct flow rates, virtually any desired concentrations of volatile and nonvolatile test solution components can be achieved. This approach also allows for the possibility of changing test solution composition in a defined way during injection. For example, the concentration of one component can be varied by varying the flow rate of the pump used to transfer the solution containing that component.

Gaseous test solutions used in vadose zone tests can also be prepared to any desired composition. In one approach, the various gases that makeup the test solution are stored in separate compressed gas cylinders, and mass flow controllers are used to combine the separate gas streams at known rates to obtain the desired composition. Alternatively, a customized gas mixture is prepared by transferring portions of each gas to an empty cylinder to achieve a mixture with the desired partial pressure of each gas. Finally, gases can be introduced into collapsible bags as described above, mixed, and then injected using a pump.

In all cases, the composition of aqueous and gaseous test solutions should be confirmed by periodic sampling during the injection phase. The potential difficulties in creating precisely controlled chemical compositions in large aqueous or gaseous volumes cannot be overemphasized. If possible, the procedures to be used to prepare test solutions in the field should first be evaluated with identical equipment, etc. in the laboratory. Particular care should be taken in confirming the ability to produce desired concentrations of highly volatile or low solubility components. During field testing, samples of the test solution should be collected from the early, middle, and late portions of the injection period because several time-variable factors can contribute to unanticipated changes in test solution composition during injection. For example, temperature may change during the day, affecting solubility, vapor pressure, etc., pump calibrations may change as water levels in tanks and wells change, or mass flow controllers may also drift out of calibration as supply gas pressures change.

2.5 Test Solution Injection

A number of factors influence the injection method including: the volume and composition of the test solution, the depth and conductivity of the aquifer, well construction details, and a number of logistical issues (number of tests, availability of pumps, power, etc.). Aqueous test solutions may be siphoned or pumped into the well or added manually ("poured") in aliquots using suitable containers. Siphons are convenient in situations where the injection times are anticipated to be large (large volumes of aqueous test solution or small formation conductivity). However, it can be difficult to control injection rates (because of varying water levels) unless in-line valves and flow meters are used but for most types of tests the exact injection rate is not needed to interpret test data. Pumps of various kinds are widely used to inject aqueous test solutions, and are usually required if precise control of injection rate is required (e.g. in tests designed to estimate regional groundwater velocity). Injection rates for aqueous test solutions can be monitored by volumetrically measuring water levels in the supply tank, weighing collapsible bags using a portable scale, or with inexpensive in-line flow meters. Gaseous test solutions are typically injected using pressure from the compressed gas cylinders and injection rates are controlled by controlling the supply pressure and measured using in-line gas flow meters.

It is important to recognize that injection rates are often limited by formation permeability and well construction details (e.g. well screen length, presence of filter pack) and the condition of the well screen (old wells often have partially clogged screens). When aqueous test solutions are injected into a well, water levels in the well will rise. Depending on the length and location of the screened interval, the thickness of the annular space between formation and well casing, and the presence of filter packs, seals, etc., rising water levels (water tables in unconfined aquifers, piezometric surfaces in confined aquifers) can result in some test solution flowing upward and outward into the vadose zone above the pre-test water level (or overflowing wells at the land surface in shallow tests). In addition to delivering a portion of the prepared test solution to a part of the formation that may not be the subject of study, some of the test solution injected into the vadose zone will be retained by capillary forces and will not be accessible for subsequent extraction and sampling, thus increasing dilution losses. A similar process can occur with gaseous test solutions. Water levels or gas pressures should be monitored during injection using suitable water level meters or pressure gages. In fact, by measuring flowrates and pressures during test solution injection, it is possible to determine the formation permeability using conventional pumping test theory and analytical methods during a push-pull test with little added effort.

Test solutions usually enter the well through tubing that extends from the surface to a point within the screened interval of the well. If the well screen is long or the well diameter is large, injected test solutions may not completely mix with the "dead volume" of fluid stored within the casing, resulting in varying concentrations of test solution components entering the formation (e.g. lower concentration higher up in the saturated zone) Incomplete mixing above ground or in the well casing is another form of "dispersion" that will contribute to increased dilution losses and "tailing" of concentration profiles, which might be confused with mass-transfer processes such as diffusion controlled sorption. In many situations, this effect is relatively small compared to dilution losses caused by regional flow, formation heterogeneity, buoyancy induced flow, etc., and may not be significant if the volume of injected test solution is much larger than the well "dead volume" (this is another reason why push-pull tests are easier to conduct in small diameter wells). If deemed important, this problem can sometimes be ameliorated by mixing the contents of the well casing during test solution injection (and perhaps during subsequent sampling). One simple way that this can be accomplished is by inserting a weighted porous stone attached to a gas supply line (like those used to mix test solutions with compressed gas) to the bottom of the well, and bubbling the contents of the well casing with a suitable compressed gas. Care in mixing the contents of the well is particularly important for certain types of push-pull tracer tests that assume solute concentrations in fluids leaving and entering the well are uniform across the formation thickness. Another push-pull test variation is to simply bubble gas mixtures in the well with no aqueous solution injection and allow regional flow and diffusion to transfer tracers from the well to the formation. This approach has the advantage that no above-ground storage or preparation of test solutions is required to introduce gaseous tracers into the formation.

Test solutions may be injected across the entire length of an uncased borehole, across the entire length of the screened interval of a well, from the tip of a drive-point, or from a multi-level well screen. In some cases it is desirable to conduct push-pull tests within particular depth intervals in uncased wells or wells with long screened intervals. Straddle packers may be used for this purpose. Straddle packers are lowered to the desired depths and inflated to isolate a portion of the borehole or well casing. The test solution is then injected and extracted from the portion of the well screen between the two packers. Dedicated multilevel wells or multiple closely-spaced wells installed at different depths may also be used for this purpose.

The total volume of test solution injected should be recorded (e.g. by recording falling water levels in the tank(s) used to prepare the test solution) as this provides an estimate of the formation volume interrogated during the test and is useful for computing mass balances. For aqueous test solutions mass balance calculations are performed by integrating measure concentrations and volumes during injection. For gaseous test solutions, these calculations are performed by integrating measured gas flow rates, pressures, and temperatures during injection.

In some cases, test solution injection is immediately followed by injection of a "chaser" solution into the well. The "chaser" typically contains no added tracers. "Chaser" injection insures that all injected test solution is pushed out of the well casing and into the formation. Although this may be required for certain types of tracer tests (i.e. those aimed at determining pore fluid velocities or mass transfer coefficients) it should not be adopted as a general practice since the injected "chaser" will also lead to increased dilution of the injected test solution.

2.6 Extraction/Sampling

After injection is completed, samples of the test solution/pore fluid mixture are collected in a variety of ways that depend on the purpose of the test. For many types of tests, the test solution is injected and then samples of the test solution/pore fluid mixture are collected as a series of discrete sampling events distributed over a defined period of time. For other types of tests, extraction pumping begins immediately after test solution injection is complete and continues at a constant rate until a pre-determined total volume has been extracted. A "rest phase" with no pumping may be included between the injection and extraction phases (e.g. to allow the injected test solution to drift with regional flow). The overall goal, of course, is to allow the test solution to reside in the formation long enough to detect the targeted process or reaction. Thus, tracer tests designed to measure fluid velocity require that the test solution reside in the formation sufficiently long to be advected downgradient from the well. Similarly, longer residence times may be required for tests aimed at determining diffusion-controlled mass transfer coefficients or small (slow) reaction rates. Of course, sampling frequency could only be optimized if the ambient fluid velocities, reaction rates etc. where known prior to the start of the test! Thus, whenever possible it is desirable to perform sample analyses while the test is in progress so that the need for continued sampling can be evaluated in "real time". Perhaps the largest uncertainty is in the dilution rate and, in general, samples collected after nonreactive tracer concentrations have fell below background levels will provide any useful information. Thus, at a minimum, it is desirable if nonreactive tracer concentrations be determined after each sampling event. When this is not possible, it is often desirable to sample more frequently and to continue sampling for a longer period of time to insure that that sampling captures the process of interest. Then, the costs of sample analyses can be optimized in the laboratory (e.g., analyzing nonreactive tracer concentrations first, analyzing "every-other" sample, etc.)

2.7 Data Analysis

Data analysis begins by preparing concentration profiles (i.e. measured concentration vs. time) for all injected tracers and any products formed in situ and, in general, these are interpreted using the same techniques used to interpret any laboratory experiment, *with one important exception*. The concentration profiles of the reactive tracers and reaction products must first be adjusted for dilution using measured concentrations of the nonreactive tracer(s). A useful analogy is to consider push-pull test data to have come from a "leaky batch reactor". Each sample collected during the extraction phase of a push-pull test is typically a blend of the injected test solution and the ambient pore fluid (which "leaks" into the test volume), i.e. the injected test solution has been "diluted" by ambient pore fluid and each sample is actually a mixture of the two fluids. For example, Fig. 2.5 shows an example concentration profile for a Br^- tracer injected at an initial concentration of 100 mg/L. With time the injected test solution migrates away from the well and the Br^- concentration decreases. By plotting the relative concentration, C/C_o, where C is a measured Br^- concentration and C_o is the Br^- concentration in the injected test solution (Fig. 2.5), it is possible to quantify the extent of dilution that has occurred. Thus, in Fig. 2.5, a sample collected 20 h after injection has a measured Br^- concentration of 36 mg/L and a relative concentration of 0.36. The latter is interpreted to mean that this sample consists of a mixture of 36 % injected test solution and 64 % ambient groundwater. As will be shown in the Examples, relative concentrations for injected nonreactive tracers are used to compute *dilution-factors*, which are then applied to measured concentrations of reactive tracers to obtain *dilution-adjusted concentration profiles* for the reactive tracers. Dilution-adjusted concentration profiles are used, for example, to compute reaction rates. Of course, several assumptions and conditions are required when applying dilution factors computed from a nonreactive tracer to a reactive tracer and these will be discussed in the Examples.

A variety of specific methods have been developed to interpret push-pull test data for different applications and these are described in detail in the Examples and the primary literature. It should be noted that several features of the push-pull test contribute to simplify data analysis. For example, push-pull tests are generally less sensitive to aquifer heterogeneity than well-to-well tests because of the flow reversal that occurs between injection and extraction phases. For this reason, concentration profiles obtained from push-pull tests tend to be relatively smooth, with simple monotonic decreases in injected tracer concentrations, etc. Nevertheless, each method of data analysis involves making certain simplifying assumptions and the validity of these should be evaluated for each specific application, as discussed in the Examples.

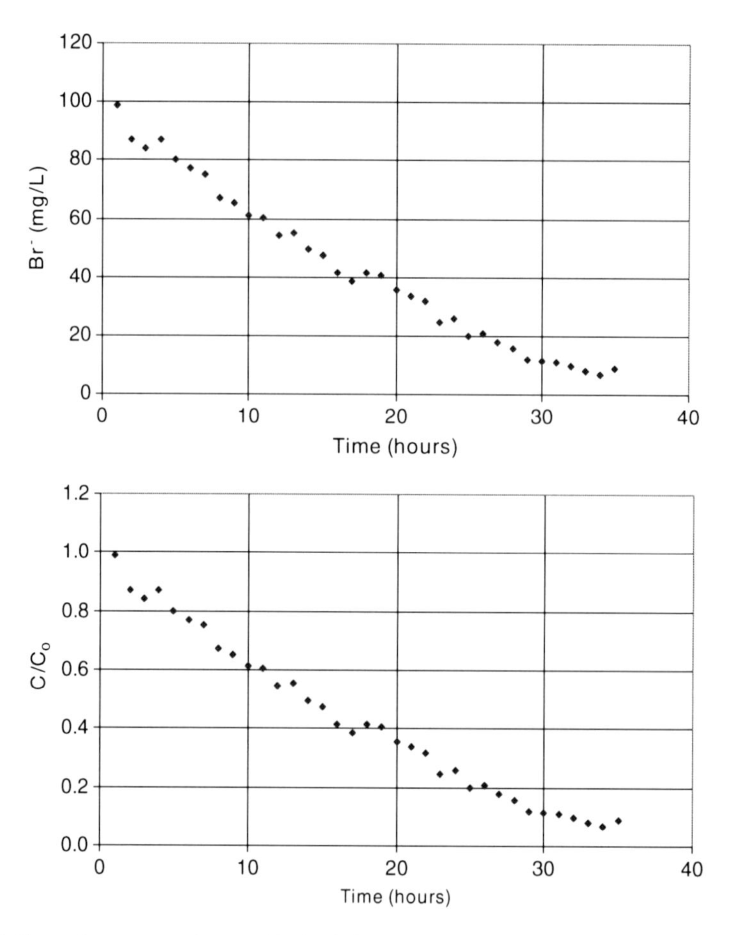

Fig. 2.5 Example concentration profile for injected Br^- as a nonreactive tracer. The relative concentration, C/C_o, where C is a measured Br^- concentration in an extraction phase sample and C_o is the Br^- concentration in the injected test solution, serves as a measure of dilution

Chapter 3
Push-Pull Test History

Research related to what we now call push-pull tests began in the late 1960s although significant development of the method has occurred since the late 1990s. A type of single-well injection/extraction test was first used in a qualitative way by Sternau et al. (1967) to study the degree of mixing of injected water with background groundwater. The theory of dispersive transport in the radial flow field near an injection well was first presented by Hoopes and Harleman (1967) who were investigating processes occurring during the recharge and disposal of injected liquid wastes and wastewater. They solved the advection-dispersion equation including linear sorption to predict tracer concentrations as a function of radial distance and time from the injection well. They also conducted a large-scale (~6 m) laboratory push-pull test in a semi-cylindrical sand box to compare experimental results with model predictions.

Until the late 1990s the method was most widely used in the oil industry to determine residual oil saturation in petroleum reservoirs (Tomich et al. 1973). *Residual oil* refers to the amount of oil remaining in the reservoir after active pumping has removed most oil from the pore space. This quantity is used for material balance calculations and to estimate the economics of attempting to extract residual oil using more aggressive removal techniques (e.g. injecting water or steam to mobilize residual oil to extraction wells). In this push-pull test application, the injected test solution contained ethyl acetate as a reactive tracer. Within the oil reservoir, a portion of the injected ethyl acetate is hydrolyzed to ethanol and the concentrations of both solutes are measured during the extraction phase. If residual oil is present in the portion of the reservoir investigated by the test, transport of ethyl acetate (which readily partitions into the stationary oil phase) is retarded relative to the more water soluble ethanol. The residual oil saturation can be computed from the differences in arrival times for the two tracers (Tomich et al. 1973). As discussed below, alternate methods for using push-pull tests to quantify nonaqueous phase liquids in the subsurface were based on the important theoretical work of Schroth et al. (2001) and included injecting partitioning tracers (Istok et al. 2002) or the use of naturally occurring in situ tracers such as radon-222 (Davis et al. 2002, 2003).

J.D. Istok, *Push-Pull Tests for Site Characterization*, Lecture Notes in Earth System Sciences 144, DOI 10.1007/978-3-642-13920-8_3,

The term "push-pull test" was apparently first introduced by Drever and McKee (1980) who used this type of test to investigate sorption characteristics of an aquifer with applications to aquifer restoration following coal gasification and uranium extraction. In their early tests they injected 15–20 m^3 of test solution and extracted ~10 times that volume to estimate parameters for linear Langmuir isotherms and cation exchange parameters using injected NH_4^+ as a sorbing reactive tracer.

Push-pull tests have been developed to measure various physical characteristics of groundwater aquifers. For example, Gelhar and Collins (1971) derived equations that can be used to determine the longitudinal dispersivity of an aquifer from the extraction phase breakthrough curve for an injected nonreactive tracer considering advection, dispersion, and diffusion. Leap and Kaplan (1988) and Hall et al. (1991) derived equations that can be used to determine the effective porosity and regional groundwater velocity from the results of a push-pull test with a rest phase between test solution injection and continuous extraction, which Hall et al. referred to as a "drift-pumpback" test because in this application injected test solutions are allowed to drift with regional groundwater flow for a period of time prior to extraction pumping.

Beauheim (1987), Haggerty et al. (2000, 2001), Meigs and Beauheim (2001), and others developed the theoretical framework and data interpretation methodology for using push-pull tests to quantify diffusion-controlled mass transfer and to estimate mass transfer coefficients, primarily in fractured dolomite with applications to a geologic waste repository. They analyzed tracer test data to estimate parameters of both dual-porosity and multiple rate diffusion models.

Push-pull tests have also been deployed to investigate a variety of other physical/chemical phenomenon. Swartz and Gschwend (1999) investigated the release of colloids from aquifer surfaces in response to varying pH, phosphate, and ascorbic acid added to injected test solutions. Field et al. (1999) conducted a series of push-pull tests to determine the effectiveness of surfactants added to injected test solutions to increase the solubility of trichloroethene in groundwater. Field et al. (2000) conducted a series of push-pull tests to characterize and quantify cation exchange processes in aquifer sediments. In that study injected Na^+ caused the release of Mg^{2+}, Ca^{2+} etc., from mineral surfaces.

Trudell et al. (1986) perhaps first applied a simple form of push-pull test to study microbial processes in the subsurface. In their study a modified drive sampler was used to inject and extract fluids from a specific depth interval in a specially prepared borehole to assay for denitrification in an alluvial aquifer. A push-pull test to study hydrocarbon degradation under denitrifying and sulfate-reducing conditions was performed by Reinhard et al. (1997). In that study, relatively large volumes (750–900 L) of test solution containing a nonreactive tracer, hydrocarbons, and added electron acceptors were injected into an existing well. Prior to injected the reactive tracers, a large volume of tracer-free water was injected to flush hydrocarbons present in background groundwater out of the portion of the aquifer near the test well.

A series of push-pull tests was performed by Istok et al. (1997) to determine rates of aerobic respiration, denitrification, sulfate reduction, and methanogenesis in a petroleum contaminated aquifer by measuring the loss of injected O_2, NO_3^-, SO_4^{2-}, H_2, and the production of CO_2, NO_2^-, and CH_4 relative to the dilution losses of a coinjected Br^- tracer. Results supported the hypothesis that petroleum contamination resulted in increased activity of indigenous microorganisms. The site-scale spatial variability in these processes was quantified in a second study at the same site (Schroth et al. 1998). A number of recent studies have utilized push-pull tests to investigate microbially mediated processes in petroleum contaminated aquifers (Schroth et al. 2001; McGuire et al. 2002; Kleikemper et al. 2002; Pombo et al. 2005; Reusseur et al. 2002; Burbery et al. 2004). In some cases, these studies have supplemented push-pull test activity measurements with a variety of ancillary data including stable isotope analyses, microbial measurements, and the use of injected deuterium- and ^{13}C labeled contaminants or other substrates as reactive tracers. The use of labeled contaminants allows sensitive detection of these compounds and their transformation products in the presence of high or variable background (unlabeled) contaminant concentrations. Thus, Reusser et al. (2002) were able to monitor anaerobic transformation of injected deuterium labeled ("deuterated") toluene and xylene to deuterated benzylsuccinic acids in a petroleum contaminated aquifer containing abundant toluene, xylene, and other hydrocarbons. Stable isotope analyses can be used to detect the diagnostic shift in isotopic composition (isotopic fractionation) that occurs as an organic compound is microbially transformed. Thus, since microorganisms preferentially utilize the "lighter" isotopes, the isotopic mixture of an injected compound becomes enriched in the "heavier" isotopes during a test.

Push-pull tests have also been performed to quantify rates of anaerobic transformations of chlorinated solvents (Hageman et al. 2001), polynuclear aromatics (Borden et al. 1989), and radionuclides (Senko et al. 2002; Istok et al. 2004). The Hageman et al. study is significant because it introduced the use of flourine labeled ("fluorinated") analogs of the common chlorinated solvents perchloroethene, trichloroethene, etc. as reactive tracers to investigate anaerobic transformations of the targeted chlorinated solvents. The fluorine analogs are very chemically similar to and are expected to undergo similar microbiological transformations as the corresponding chlorinated solvents but are not present in groundwater (even in groundwater contaminated with chlorinated solvents). This makes it possible to detect and quantify reactions affecting chlorinated solvents by injecting the fluorinated analogs during push-pull tests. Thus, Hageman et al. injected test solutions containing the fluorinated analog compound trichloro-fluoroethene (TCFE) as a reactive tracer to determine rates of reductive dechlorination for the targeted chlorinated solvent trichloroethene (TCE). The use of TCFE is similar to the use of deuterium or ^{13}C-labeled compounds, except that the fluorinated compounds are much less expensive and provide a more sensitive chemical signature when background contaminant concentrations are high. Thus Hageman et al. was able to monitor TCFE transformations in a TCE contaminated aquifer even when background TCE concentrations were as high as 400 mg/L. Additional details on the use of TCFE as a reactive tracer are in Field et al. (2005); the use of chlorofluroethene (CFE) as a tracer for vinyl chloride is in Ennis et al. (2005).

Kim et al. (2004) and Kim et al. (2006) conducted many push-pull tests to detect and quantify rates of aerobic cometabolism of TCE and cis-dichloroethene (cis-DCE). Injected test solutions contained varying concentrations of Br^- and Cl^- as nonreactive tracers, propane, oxygen, and nitrate to stimulate microbial growth, and ethylene and propylene to detect and quantify the cometabolic activity of indigenous microorganisms. Ethylene and propylene are additional examples of analog reactive tracers. Both compounds are transformed by the same monogenase enzyme systems responsible for TCE and cis-DCE transformation but are not present in the subsurface, even at contaminated sites. Thus, these compounds can be injected to probe the activity of indigenous microorganisms even in the presence of background TCE etc. Stable isotope labeled reactive tracers (e.g. deuterium and N^{15} labeled compounds) have also been used in push-pull tests (Pombo et al. 2002; Reusseur et al. 2002; Schürmann et al. 2003).

Several of these studies have investigated the effects of various amendments on rates of microbial activity or contaminant transformation. This has consisted of either a series of tests conducted in the same well or in several nearby wells, some tests with added substrates to increase rates of microbial activity and other tests without added substrates for comparison or to serve as controls. Thus, Kim et al. (2006) stimulated growth with injected propane and oxygen and also inhibited aerobic cometabolism with injected acetylene in some tests. Hageman et al. (2001) measured rates of reductive dechlorination with and without added formate or lactate, Reinhard et al. (1997) measured rates of petroleum transformation with and without added nitrate and sulfate, and Istok et al. (2004) measured rates of uranium and technetium reduction with and without added ethanol, acetate, or glucose to increase the size of targeted groups of microorganisms. Istok et al. (2004) also used injected acetylene to inhibit denitrification in some tests. Gu et al. (2005) added natural humic compounds to accelerate rates of electron transfer during uranium reduction.

Although most push-pull test applications to date have involved tests in monitoring wells in groundwater aquifers the potential scope of push-pull test applications is much larger. For example, Luthy et al. (2000) performed push-pull tests to measure rates of sulfide consumption in a deep lake. In that study injection and extraction tubing were suspended in the lake water column using a buoy and anchors to hold the end of the tubing at a constant depth. Push-pull tests were possible in this system because of the weak turbulent mixing that occurred far below the lake surface.

Although most push-pull tests have been conducted using aqueous test solutions, a few studies have used gaseous test solutions to interrogate the unsaturated zone (Urmann et al. 2005; Gonzalez-Gil et al. 2006; Gomez et al. 2008). For example, Gonzalez-Gil et al. (2006) injected gas mixtures containing various noble gases as nonreactive tracers and methane and oxygen as reactive tracers to measure rates of methane oxidation in hydrocarbon contaminated sediments.

Chapter 4
Applications and Examples

4.1 Groundwater Velocity/Effective Porosity

Leap and Kaplan (1988) and Hall et al. (1991) presented a type of push-pull test for determining regional groundwater velocity and effective porosity if the hydraulic conductivity of the aquifer and local the hydraulic gradient are known. The hydraulic conductivity could be determined by a pumping test conducted in the same well at the same time as the push-pull tracer test described here; the hydraulic gradient could be determined from water level measurements in a set of nearby wells surrounding and including the push-pull test well. The procedure involves injecting a constant-concentration test solution containing a nonreactive tracer into the aquifer using a single well, allowing the test solution to drift downgradient with the regional groundwater flow, and then extracting the tracer solution/groundwater mixture from the same well by continuous pumping to determine the temporal displacement of the tracer center of mass. The basic equations (using the notation of Hall et al. 1991) are:

$$q = \frac{Qt}{\pi b d^2 K I} \tag{4.1}$$

$$n = \frac{\pi b K^2 I^2 d^2}{Qt} \tag{4.2}$$

where q is the apparent groundwater (Darcy) velocity, n is effective porosity, Q is the extraction pumping rate, t is the time elapsed from the start of extraction pumping until the centroid of the tracer mass has been extracted, b is the aquifer saturated thickness, d is the elapsed time from the end of tracer injection until the centroid of the tracer mass is extracted (drift time + t), K is the saturated hydraulic conductivity, and I is the local hydraulic gradient. Obviously, uncertainties in computed values of q and n will reflect uncertainties in values of K and I.

J.D. Istok, *Push-Pull Tests for Site Characterization*, Lecture Notes in Earth System Sciences 144, DOI 10.1007/978-3-642-13920-8_4,

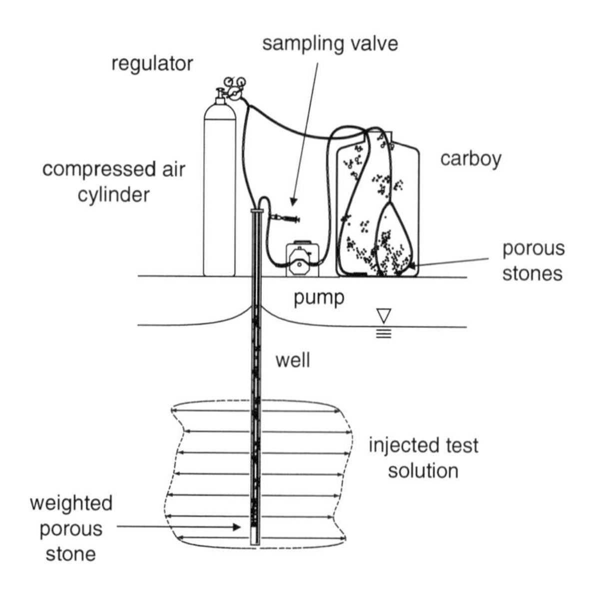

Fig. 4.1 Equipment used for push-pull test to determine groundwater velocity and effective porosity for Example 1.0

For this example, a push-pull test was conducted in a shallow, unconfined aquifer formed in an alluvial sand/gravel deposit. A 5 cm diameter monitoring well was selected for testing. The well is 3.5 m deep and the initial water-table depth was 1 m. Pumping tests conducted in this well gave a value of saturated hydraulic conductivity of 2.4 m/day; the local hydraulic gradient determined from measured water levels in this well and two nearby wells was 0.015 prior to the start of the test. A peristaltic pump was used to collect 200 L of groundwater from the well in a plastic tank at a rate of ~2 L/min. An additional 50 L of groundwater was collected in a second plastic tank for use as a tracer-free "chaser". Sufficient KBr was added to the water in the 200 L tank to achieve a Br^- concentration of ~100 mg/L and the test solution was mixed overnight using a tank of compressed air to deliver air to three porous stones placed on the bottom of the tank (Fig. 4.1). No KBr was added to the "chaser" water. The injection of tracer-free "chaser" is generally not advisable as it simply increases the dilution of the injection test solution but in this type of test it is convenient because it helps to more clearly delineate the center of mass of the tracer pulse during the extraction phase.

After the water level in the well had recovered from makeup water collection, the Br^- tracer test solution was injected at a constant rate of 2 L/min using a peristaltic pump. A timer was started when injection began and elapsed time from this moment was used to record all subsequent test events. During the test, a weighted porous stone was placed in the bottom of the well and connected to a compressed air cylinder to mix the contents of the well casing (Fig. 4.1). Test solution injection resulted in an increase in water table elevation of ~15 cm. Five samples of the test solution were collected in 8 mL glass screw-top vials using a sampling valve placed in the injection line on the positive pressure side of the

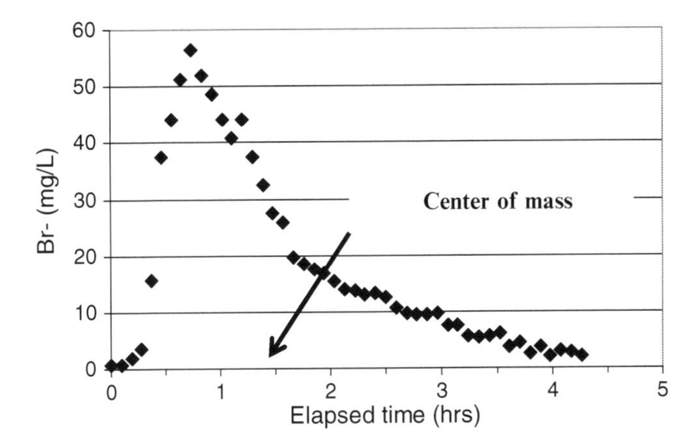

Fig. 4.2 Measured Br^- concentrations during extraction phase of push-pull test to determine groundwater velocity and effective porosity for Example 1.0

peristaltic pump. These were collected when the water level in the tank read 200, 150, 100, 50, and 5 L. Tracer injection was completed in 100 min and was immediately followed by the injection of 50 L tracer-free "chaser", also at 2 L/min for another 25 min. After test solution and "chaser" were injected pumping ceased; extraction pumping did not begin until 30 h after the end of test solution injection. During this "rest" phase, the test solution gradually drifted downgradient with the regional groundwater flow. Then, the test solution/groundwater mixture was pumped from the well using a peristaltic pump at 2 L/min. The volume of extracted water was recorded by filling calibrated buckets and water samples were collected from the pump discharge line after each 10 L had been extracted. The elapsed time and cumulative extraction volume were recorded for each sample. Periodic measurements of Br^- concentration were made in the field using an ion specific electrode and meter and extraction pumping continued until measured Br^- concentrations had decreased to ~1 mg/L, which occurred after 460 L had been extracted. All samples were subsequently analyzed for Br^- by ion chromatography.

Br^- concentrations during the test are summarized in Fig. 4.2. The background (pre-test) Br^- concentration was <0.5 mg/L, supporting the use of Br^- as a tracer. The average Br^- concentration in the injected test solution was 100 mg/L. Initially, Br^- concentrations during the extraction phase were small (due to the "chaser" injection and to the drift of the test solution away from the well with the regional groundwater flow). Then Br^- concentrations increased to a maximum and then gradually decreased and approached the background Br^- concentration. Using these data, the time to the center of mass of the Br^- breakthrough curve was determined to have occurred 1.45 h after the start of the extraction phase. The time to the center of mass for this test, the parameters in Eqs. 4.1 and 4.2 are: Q = 1.8 L/min = 0.108 m^3/h; t = 1.45 h; b = 2.93 m, d = 30 + 1.45 = 31.45 h, K = 2.4 m/day = 0.10 m/h, and I = 0.015:

$$q = \frac{\left(0.108\ \frac{m^3}{h}\right)(1.45\,h)}{(3.14)(2.93\,m)(31.45\,h)^2\left(0.10\ \frac{m}{h}\right)(0.015)} = 0.278\,\frac{m}{day}$$

$$n = \frac{(3.14)(2.93\ m)\left(0.1\,\frac{m}{h}\right)^2(0.015)^2(31.45\,h)^2}{\left(0.108\,\frac{m^3}{h}\right)(1.45\,h)} = 0.28$$

4.2 Dispersivity

Methods for estimating aquifer longitudinal dispersivity from push-pull test data were developed by Mercado (1966), Gelhar and Collins (1971), Pickens and Grisak (1981), and Schroth et al. (2001). The governing equation for one-dimensional (radial) transport of a nonreactive tracer in a homogeneous, isotropic aquifer with constant saturated thickness is:

$$\frac{\partial C}{\partial t} = \alpha_L v \frac{\partial^2 C}{\partial r^2} - v \frac{\partial C}{\partial r} \tag{4.3}$$

where C is tracer concentration, t is time, α_L is longitudinal dispersivity, v is the average porewater velocity, and r is radial distance. For a push-pull test, porewater velocity varies as a function of radial distance from the pumping well during both injection and extraction phases:

$$v(r) = \frac{Q}{2\pi bnr} \tag{4.4}$$

where Q is the pumping rate (positive for injection and negative for extraction), b is the saturated aquifer thickness, and n is the effective porosity. Equations 4.3 and 4.4 assume that the regional groundwater velocity is negligible compared to the velocity field imposed by pumping and that molecular diffusion can be ignored. This assumption is generally valid because pumping at the test well typically dominates the local hydraulic gradient but can be checked if the regional groundwater velocity is known (e.g. by performing the type of push-pull tracer test described in Example 1). Gelhar and Collins (1971) derived approximate solutions to Eq. 4.3 for the case of constant concentration tracer injection into a tracer-free aquifer. The solution for relative tracer concentration during the extraction phase is

$$\frac{C}{C_o} = \frac{1}{2}\operatorname{erfc}\left[\frac{\left(\frac{V_{ext}}{V_{inj}} - 1\right)}{\left[\left(\frac{16\alpha_L}{3r_{max}}\right)\left(2 - \left[1 - \frac{V_{ext}}{V_{inj}}\right]^{1/2}\left(1 - \frac{V_{ext}}{V_{inj}}\right)\right)\right]^{1/2}}\right] \tag{4.5}$$

where C_o is the concentration of the injected tracer, V_{ext} is the cumulative extracted volume (i.e., the total volume of water removed during the extraction phase of the test), V_{inj} is the volume of injected tracer solution, and r_{max} is the "maximum frontal position" given by

$$r_{max} = \left[\frac{V_{inj}}{\pi bn} + r_w^2\right]^{1/2} \tag{4.6}$$

where r_w is the well radius and is the distance the tracer would be expected to penetrate into the aquifer if only advective transport was occurring. Dispersivity is estimated by fitting Eq. 4.5 to measured tracer concentrations during the extraction phase.

For this example, a push-pull tracer test was conducted in a shallow, unconfined aquifer formed in an alluvial sand/gravel deposit in the same monitoring well used in Example 1. The well is 5 cm in diameter and 3.5 m deep with an initial water-table depth of 1 m. The effective porosity of the aquifer was determined in Example 1 to be 0.28. A peristaltic pump was used to collect 200 L of groundwater from the well in a plastic tank at a rate of 2 L/min. Sufficient KBr was added to the water in the 200 L tank to achieve a Br^- concentration of ~100 mg/L and the test solution was mixed overnight using compressed air to deliver air to three porous stones placed on the bottom of the tank.

After the water level in the well had recovered from makeup water collection, the Br^- tracer test solution was injected at a constant rate of 2 L/min using a peristaltic pump. A timer was started when injection began and elapsed time from this moment was used to record all subsequent test events. During the test, a weighted porous stone was placed in the bottom of the well and connected to compressed air to mix the contents of the well casing as in Example 1. Five samples of the test solution were collected during injection; the average Br^- concentration in the injected test solution was 98.8 mg/L. Immediately after injection was completed, the test solution/groundwater mixture was extracted from the well using a peristaltic pump at 2 L/min. The volume of extracted water was recorded by filling calibrated buckets and water samples were collected after each 14 L had been extracted. The cumulative extraction volume, V_{ext} was recorded for each sample. Extraction pumping continued until ~490 L had been extracted. All samples were analyzed for Br^- by ion chromatography.

Br^- concentrations during the extraction phase are plotted as relative concentrations, C/C_o, where C is the measured Br^- concentration in a sample and C_o is the average Br^- concentration in the injected test solution, versus the

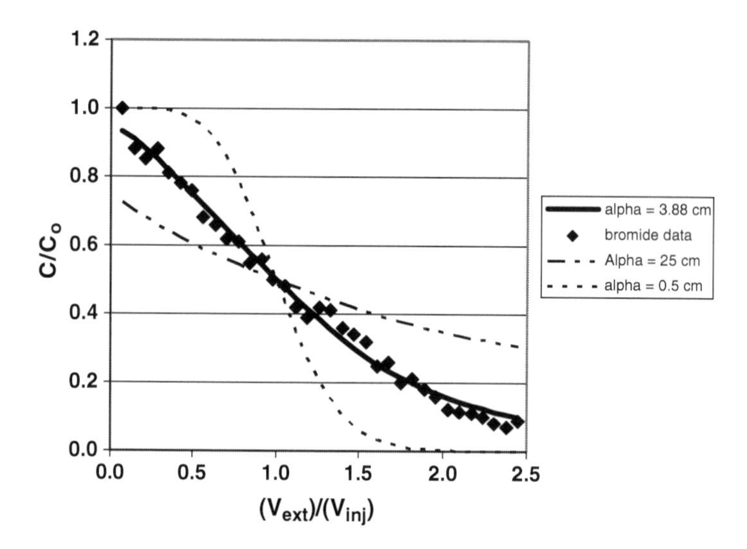

Fig. 4.3 Measured and fitted Br^- concentrations during push-pull tracer test. Alpha = α_L. Also shown are simulated Br^- concentrations for a smaller and larger dispersivities

ratio V_{ext}/V_{inj} (Fig. 4.3). For this test, the parameters in Eq. 4.6 are V_{inj} = 200 L = 2×10^5 cm^3; b = 2.5 m = 250 cm; n = 0.27; r_w = 2.54 cm and

$$r_{\max} = \left[\frac{2 \times 10^5 \mathrm{cm}^3}{\pi(250\,\mathrm{cm})(0.28)} + (2.54\,\mathrm{cm})^2\right]^{1/2} = 30.2\,\mathrm{cm}$$

The value of r_{max} represents the maximum radial distance that the injected tracer penetrates into the aquifer by advection only at the end of the injection phase (i.e., the maximum radial distance where $C/C_o = 0.5$) assuming homogeneous conditions, uniform geometry, etc. Equation 4.5 was fit to the experimental data by varying the value of α_L until a minimum sum of squared errors between measured and predicted Br^- concentration was obtained (Fig. 4.3). The best-fit value was α_L = 3.88 cm. Of course, dispersivity is related to the effective pore-size distribution, which is known to be heterogeneous and to vary with tracer transport distance, which in this case is determined by the test solution injection volume, V_{inj}. In this test, the injection volume was relatively small and the tracer likely only penetrated a relatively small distance into the formation and the estimated value of α_L therefore reflects near-well conditions and likely does not represent site-scale tracer behavior. An improved estimate could be obtained by repeating the test with an increased volume of injected tracer solution. It should also be noted that this type of test can also be performed using a 'resident' nonreactive tracer (e.g. Cl^-) instead of an injected tracer. This can be convenient when injection volumes are large or when regulatory approval cannot be obtained for tracer injection. In this case, tracer-free water is injected and concentrations of the resident tracer are made during the extraction phase. The symmetry of the governing equations allows the resulting breakthrough curve (plotted inversely as $1-C/C_b$, where C_b is the pre-test concentration of the resident tracer) to be fit to obtain an estimate for dispersivity.

4.3 Retardation Factors

Schroth et al. (2001) developed a method for estimating retardation factors from push-pull tests that use test solutions containing a nonsorbing nonreactive tracer and a second potentially sorbing but nonreactive tracer. The method assumes an equilibrium linear sorption isotherm

$$S = K_d C \tag{4.7}$$

where S and C are the sorbed and aqueous concentrations for the sorbing tracer, respectively and K_d is a distribution coefficient. Using Eq. 4.7, the retardation factor, R is defined as

$$R = 1 + \frac{\rho_b}{n} K_d \tag{4.8}$$

where ρ_b and n are the bulk density and porosity of the aquifer, respectively ($K_d = 0$ and $R = 1$ for the nonsorbing tracer). The governing transport equation for one-dimensional (radial) flow is

$$R \frac{\partial C}{\partial t} = \alpha_L v \frac{\partial^2 C}{\partial r^2} - v \frac{\partial C}{\partial r} \tag{4.9}$$

where C is the aqueous tracer concentration, t is time, α_L is longitudinal dispersivity, v is the average porewater velocity, and r is radial distance. Sorption retards solute transport; the effective velocity, v* for a sorbing tracer is a function of radial distance

$$v^*(r) = \frac{v}{R} = \frac{Q}{2\pi bnrR} \tag{4.10}$$

where Q is the pumping rate (positive for injection and negative for extraction), b is the saturated aquifer thickness, and n is the effective porosity. Equations 4.7 and 4.8 assume that the regional groundwater velocity is negligible compared to the velocity field imposed by pumping and that molecular diffusion is negligible. Gelhar and Collins (1971) derived approximate solutions to Eq. 4.7 for the case of constant concentration tracer injection into a tracer-free aquifer. The approximate solution for tracer concentration during the extraction phase is

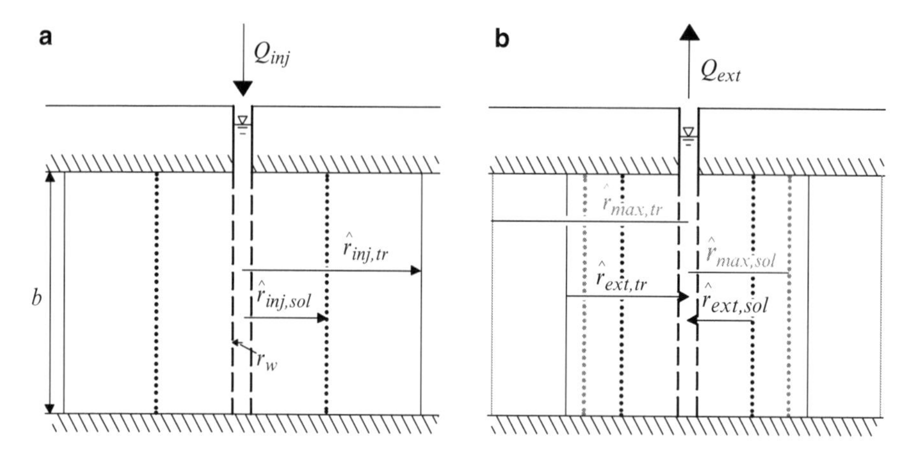

Fig. 4.4 Maximum frontal positions for a nonsorbing tracer (subscript "tr") and a co-injected sorbing tracer (subscript "sol") during (**a**) the injection phase and (**b**) prior to (*gray lines* and labels) and during the extraction phase of a push-pull test conducted under ideal transport conditions in a homogeneous confined aquifer (Schroth et al. 2001)

$$C = \frac{C_o}{2}\,\mathrm{erfc}\left[\frac{\left(\frac{V_{ext}}{V_{inj}} - 1\right)}{\left[\left(\frac{16\,\alpha_L}{3\hat{r}_{max}}\right)\left(2 - \left|1 - \frac{V_{ext}}{V_{inj}}\right|^{1/2}\left(1 - \frac{V_{ext}}{V_{inj}}\right)\right)\right]^{1/2}}\right] \tag{4.11}$$

where C_o is the concentration of the injected tracer, V_{ext} is the cumulative extracted volume, V_{inj} is the injected volume, and the maximum frontal position, $\hat{r}_{max}$ is given by

$$\hat{r}_{max} = \left[\frac{V_{inj}}{\pi bnR} + r_w^2\right]^{1/2} \tag{4.12}$$

where r_w is the well radius. Note that $\hat{r}_{max}$for the sorbing tracer is always less than $\hat{r}_{max}$for the nonsorbing tracer: $\hat{r}_{max,sol} < \hat{r}_{max,tr}$(Fig. 4.4).

Under assumed ideal transport conditions (i.e., homogeneous and isotropic aquifer, constant pumping rates, etc.) the longitudinal dispersivity, α_L is a property of the aquifer and should be identical for sorbing and nonsorbing tracers. Thus, differences in the extraction phase breakthrough curves for an injected sorbing tracer and a coinjected nonsorbing tracer contain information about the sorption process and can be used to estimate retardation factors for the sorbing tracer. It is assumed that sorption alone accounts for the differences in breakthrough curves for the two tracers. The effect of sorption is to create greater apparent dispersion in the extraction phase breakthrough curve for the sorbing tracer compared to the nonsorbing tracer (Fig. 4.5). The procedure of Schroth et al. (2001) consists of a series of steps:

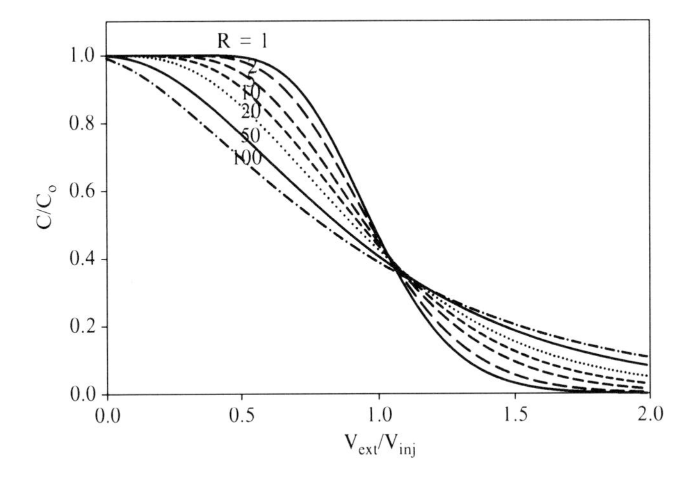

Fig. 4.5 Simulated breakthrough curves for a nonsorbing tracer (R = 1) and co-injected nonsorbing tracers (R > 1) obtained at the injection/extraction well during the extraction phase of a push pull test conducted under ideal transport conditions illustrating the increase in apparent dispersivity with increasing retardation factor, R (Schroth et al. 2001)

1. Inject a prepared test solution containing both a nonsorbing tracer ($K_d = 0$; $R = 1$) and a sorbing tracer ($K_d > 1$; $R > 1$).
2. Extract the test solution and periodically measure tracer concentrations as a function of time to prepare breakthrough curves for each tracer. Breakthrough curves plot C/C_o versus V_{ext}/V_{inj} for each tracer in a form that may be fit by Eq. 4.11.
3. Compute the maximum frontal position for the nonsorbing tracer, $r_{max,tr}$ using

$$\hat{r}_{max,tr} = \left[\frac{V_{inj}}{\pi bn} + r_w^2\right]^{1/2} \tag{4.13}$$

4. Fit Eq. 4.11 to the nonsorbing tracer breakthrough curve to obtain an estimate for α_L as in Example 2.
5. Keeping α_L fixed, estimate the maximum frontal position for the sorbing tracer, $\hat{r}_{max,sol}$, by fitting Eq. 4.11 to the breakthrough curve for the sorbing tracer.
6. Compute the estimate of the retardation factor, R* for the sorbing tracer using

$$R^* = \left(\frac{\hat{r}_{max,tr}}{\hat{r}_{max,sol}}\right)^2 \tag{4.14}$$

Schroth et al. (2001) applied this method to a push-pull test conducted by Pickens and Grisak (1981) aimed at quantifying the sorption of injected ^{85}Sr in a sandy aquifer using coinjected ^{131}I as a nonsorbing tracer. The test was conducted in an ~8 m thick layer with an average porosity of 0.38 and bulk density of 1.7 g/cm^3. The injection/extraction well was constructed of 10.4-cm-diameter

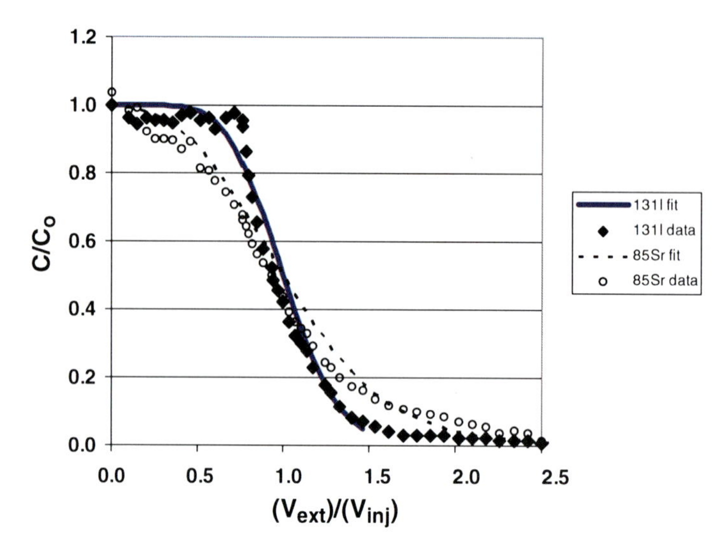

Fig. 4.6 Extraction phase breakthrough curve for coinjected nonsorbing tracer (^{131}I) and sorbing tracer (^{85}Sr) showing fits (Eq. 4.11) used to estimate α_L, $\hat{r}_{max,sol}$, R, and K_d

PVC pipe and screened across the entire layer. A volume of $V_{inj} = 244\ m^3$ of test solution containing ^{131}I and ^{85}Sr was injected, resulting in a maximum frontal position for the nonsorbing tracer (^{131}I) of

$$\hat{r}_{max,tr} = \left[\frac{V_{inj}}{\pi bnR} + r_w^2\right]^{1/2} = \left[\frac{2.44 \times 10^6\ cm^3}{(3.14)(800\ cm)(0.38)(1)} + (5.25\ cm)^2\right] = 500\ cm$$

Extraction pumping began immediately following the end of the injection phase at $Q_{ext} = 2.282\ m^3/h$ and continued until $V_{ext}/V_{inj} = 2.5$. The extraction phase breakthrough curves show the characteristic increase in apparent dispersion for the sorbing tracer (^{85}Sr) compared to the nonsorbing tracer (^{131}I) (Fig. 4.6).

Using the procedure outlined above, Eq. 4.11 was fit to the ^{131}I breakthrough curve resulting in a value of longitudinal dispersivity, $\alpha_L = 6.4$ cm (Fig. 4.6). Then, keeping α_L fixed, Eq. 4.11 was fit to the ^{85}Sr breakthrough curve resulting in a value of maximum frontal position for ^{85}Sr, $r_{max,sol} = 148$ cm (Fig. 4.6). Using Eq. 4.14 we obtain

$$R^* = \left(\frac{\hat{r}_{max,tr}}{\hat{r}_{max,sol}}\right)^2 = \left(\frac{500\ cm}{148\ cm}\right)^2 = 11.4$$

Rearranging Eq. 4.8 we obtain

$$K_d = \frac{n}{\rho_b}(R - 1) = (0.38)\left(\frac{cm^3}{1.7g}\right)(11.4 - 1) = 2.3\frac{cm^3}{g}$$

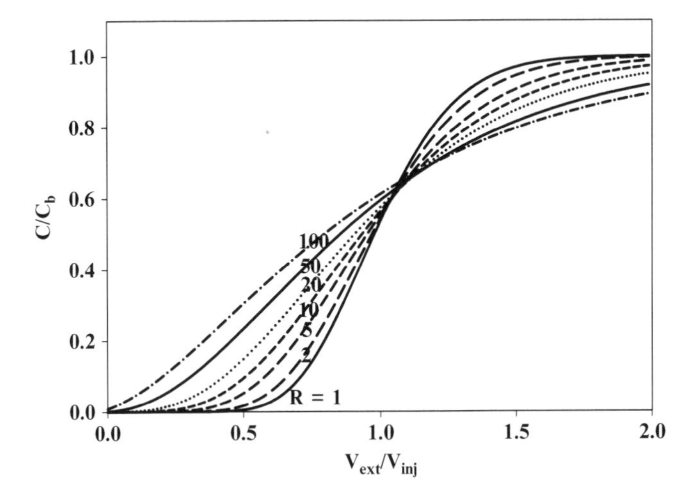

Fig. 4.7 Simulated breakthrough curves for nonsorbing tracer (R = 1) and sorbing solutes (R > 1) present in the ambient porefluid obtained at the injection/extraction well during the extraction phase of a push pull test conducted by injecting tracer-free test solution

These values are close to those obtained from field well-to-well tracer tests at this site (Schroth et al. 2001).

An interesting feature of push-pull tests conducted to quantify solute sorption behavior is that it is possible to conduct "inverse" tests if the potentially sorbing tracer is already present in the ambient ("background") porefluid. For example, it might be of interest to determine the transport behavior of a suite of organic or inorganic contaminants present in the groundwater at a hazardous waste site. If test solutions were prepared using site groundwater as the makeup water, it would be possible to conduct a push-pull test similar to the one just described except that it would be necessary to use labeled contaminants as sorbing tracers (e.g. isotopically labeled contaminants) in the injected test solution to distinguish between injected tracers and ambient contaminants. However this approach could cause logistical difficulties or increase test costs (e.g. many isotopically labeled compounds are expensive and analytical costs may also be higher). However, an alternate approach is to recognize the inherent symmetry in the extraction phase breakthrough curves for injected and ambient tracers. Thus, it is possible to conduct a push-pull test to determine retardation factors for ambient sorbing tracers only be injecting tracer free water (or perhaps water containing only a nonsorbing tracer). This is a powerful technique as the transport characteristics of many ambient sorbing tracers can be determined in a single test without preparing test solutions containing multiple sorbing tracers. In this type of test the expected extraction phase breakthrough curves for sorbing tracers with varying retardation factors are the "inverse" of those for injected sorbing tracers (Fig. 4.5) as shown in Fig. 4.7.

Of course, sorption may be dependent upon the chemical composition of the test solution (e.g. pH, ionic strength, HCO_3^- concentration, etc.). Thus, the chemical composition of the makeup water (or gas if a gaseous test solution is to be used) should be selected to insure that the transport behavior of ambient tracers is evaluated under suitable conditions for the purpose of the investigation. Recall that the chemical compositions of samples collected during the extraction phase of a push-pull test represent a mixture of injected test solution and ambient pore fluids and a determination must be made if these changes are significant to the interpretation of test results.

A push-pull test was conducted in a shallow alluvial aquifer at a uranium mill tailings reclamation area near Rifle, Co to determine the retardation factor for uranium required for modeling uranium transport at the site. Several existing monitoring wells were available for testing; all shared similar major ion chemistry, pH, and HCO_3^- content, which are considered the important factors in controlling the chemical speciation and mobility of uranium in aerobic groundwater. Groundwater from a monitoring well located in an uncontaminated portion of the aquifer was used as makeup water. 100 mg/L Br^- (as KBr) was added to serve as a nonsorbing tracer and the test solution was mixed using compressed air. 143 L of test solution were injected at 1 L/min; the average Br^- concentration in the injected test solution was 99.5 mg/L and the uranium concentration was <1 μg/L. Extraction pumping began immediately after injection was completed and continued at a rate of 1.2 L/min until 413 L had been extracted ($V_{ext}/V_{inj} = 2.89$). Samples were collected during the extraction phase and analyzed for Br^- by ion chromatography and uranium by kinetic phosphorescence analyzer. Extraction phase breakthrough curves showed a gradual decrease in Br^- relative concentrations and a gradual increase in uranium relative concentrations during the test (Fig. 4.8). In Fig. 4.8 uranium concentrations are plotted as C/C_b, where C_b is the background uranium concentration measured in the well prior to the start of the test.

The Br^- breakthrough curve was fit using the method described in Example 2 to obtain $\alpha_L = 1.94$ cm. The computed value of $\hat{r}_{max,tr}$ was

$$\hat{r}_{max,tr} = \left[\frac{V_{inj}}{\pi bn} + r_w^2\right]^{1/2} = \left[\frac{143,000\,\text{cm}^3}{(3.14)(500\,\text{cm})(0.25)} + (2.54\,\text{cm})^2\right] = 19.26\text{cm}$$

Uranium concentrations increased during the test because no uranium was present in the injected test solution but uranium was present in the ambient groundwater. The uranium breakthrough curve was fit by plotting the quantity: $1-C/C_b$ (Fig. 4.8) using the method described in Example 3 to obtain $\hat{r}_{max,sol} = 5$ cm and the retardation factor for uranium in this aquifer was computed to be

$$R^* = \left(\frac{\hat{r}_{max,tr}}{\hat{r}_{max,sol}}\right)^2 = \left(\frac{19.26\,\text{cm}}{5\,\text{cm}}\right)^2 = 14.8$$

Many variations on these types of tests are possible. For example, retardation factors have been sued to represent many simple partitioning processes and push-pull tests have been used to determine e.g. cation exchange parameters of aquifer

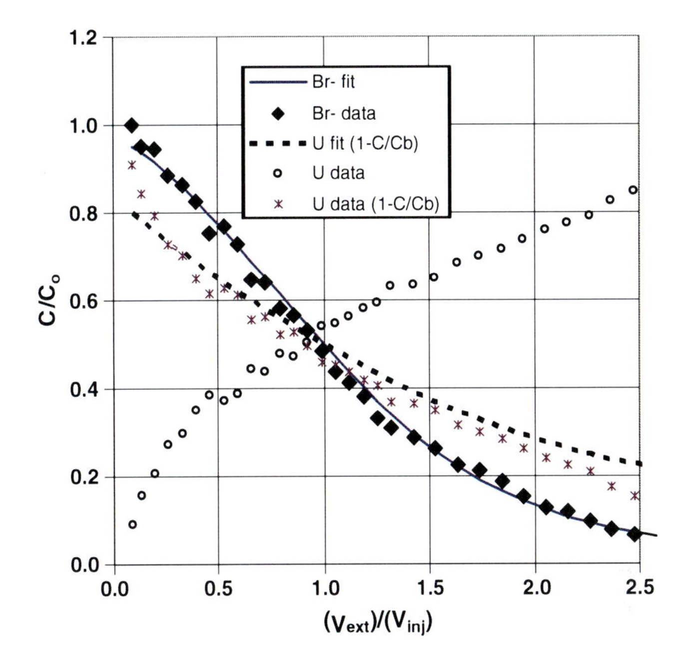

Fig. 4.8 Extraction phase breakthrough curve for Br^- (injected) and uranium (not injected but present in ambient groundwater) showing model fits used to estimate α_L and the uranium retardation factor

sediments (Drever and McKee 1980; Field et al. 2000) and to quantify residual nonaqueous phase liquid saturation (Istok et al. 1999; Davis et al. 2002, 2005). Many other applications are also possible. For example, Istok et al. (1999) used push-pull partitioning tracer tests to quantify the amount of nonaqueous phase liquid removed by surfactant flushing. Sequences of push-pull tests with varying test solution compositions also makes ti possible to determine retardation factors as a function of other system variables. For example, sorption of many metals depends on pH and on the concentration of complexing ligands (e.g. bicarbonate) and a sequence of tests with varying pH and bicarbonate concentrations can be used to parameterize a surface complexation model for metal absorption onto mineral surfaces. It is also possible to use a sequence of push-pull tests with varying test solutions as a way to quantify the fraction of sorbed solute associated with various surface or mineral phases (e.g. with increasing concentration of complexing ligand like bicarbonate or with decreasing pH).

4.4 Reaction Rates

Push-pull tests are particularly useful for determining rates of chemical and microbial reactions. In this application, test solutions are prepared with known concentrations of a nonreactive tracer and one or more potentially reactive tracers. After injection, the progress of the reaction is monitored by periodically sampling the composition of the test solution/pore fluid mixture by sampling during continuous or periodic extraction pumping. Reactions are detected and quantified by the decrease in dilution-adjusted concentrations of an injected reactive tracer or by the increase in concentrations of a reaction product formed in situ. Advection, dispersion, diffusion, sortion, mass transfer, etc., will also affect measured tracer concentrations and these are accounted for in various ways depending on the specific type of test.

The theoretical basis for estimating in situ reaction rates from push-pull test data was presented by Schroth and Istok et al. (2004). Analysis of push-pull test concentration profiles is complicated by the radial flow field and by the flow reversal between the injection (radially divergent flow) and extraction (radially convergent flow) phases of a test. Two simplified methods have been presented to determine zero-order reaction rates (Snodgrass and Kitanidis 1998) or first-order rate coefficients (Haggerty et al. 1998; Snodgrass and Kitanidis 1998). In this context, "simplified" signifies that rate estimates are determined solely from an analysis of measured tracer and reaction product concentrations, without the need for estimates of aquifer physical properties such as hydraulic conductivity, dispersivity, or porosity. Both simplified methods share two major assumptions: (1) nonreactive and reactive tracers exhibit identical transport behavior (e.g. identical advection, dispersion, diffusion, sorption behavior) in the aquifer, and (2) complete and instantaneous mixing of the injected test solution occurs in the portion of the aquifer investigated by the test, i.e., the system can be modeled as a well-mixed reactor. An alternative method for computing reaction rates from push-pull test concentration profiles when the first assumption is violated was presented by Hageman et al. (2003). On the other hand, violation of the second assumption resulted in relatively small inaccuracies in rate estimates (errors $<10\ \%$) during a sensitivity analysis performed by Haggerty et al. (1998). Nevertheless, those authors noted that rate estimates obtained with the simplified methods were generally larger than 'true' (input simulation) rates.

A 'well-mixed' reactor model for interpreting push-pull test data to obtain reaction rates was presented by Haggerty et al. (1998). Two alternative methods of data interpretation based on different mixing assumptions, the 'plug-flow' and the 'variably-mixed' reactor models, were presented by Schroth and Istok (2004).

4.4.1 Well-Mixed Reactor Model

Assuming complete and instantaneous mixing of the injected test solution in the portion of the aquifer investigated by the test, Haggerty et al. (1998) presented a simplified method for estimating first-order reaction rate coefficients, k $[T^{-1}]$ from concentration profiles for an injected reactive tracer. The reaction equation is:

$$\frac{dC_r}{dt} = -kC_r \tag{4.15}$$

where C_r is reactive tracer concentration and t is time. To illustrate the well-mixed reactor assumption, the injected test solution is imagined to be composed of several individual parcels i = 1 to n (shown in different grey shades, Fig. 4.9a). Assuming a well-mixed reactor, the system retains no memory of the parcels' injection sequence, as indicated by a single "average" grey shade for parcels j = 1 to m of test solution/ground water mixture collected during the extraction phase (Fig. 4.9b). In Fig. 4.9, y-axes show relative concentration, $C^* = C/C_o$, i.e., concentration C measured during extraction divided by the concentration in the injected test solution, C_o. Note that dilution of the injected test solution by ambient ground water, which causes the decline in C^* over the course of the extraction (Fig. 4.9b–d), is not affected by this or the other mixing assumptions. Dilution is separately accounted for during the subsequent computation of k using the nonreactive tracer data. Also note that in all analyses we assume that solute transport due to regional ground water flow is negligible compared to solute transport due to pumping during a test, and that mixing is a result of mechanical dispersion only.

For a well-mixed reactor the relative reactant concentration at any time *t* can be computed using (e.g. Jury and Roth 1990)

$$C_r^*(t) = C_{tr}^*(t)\, e^{-kt} \tag{4.16}$$

where subscripts *r* and *tr* denote reactive tracer and nonreactive tracer, respectively. Haggerty et al. (1998) also assumed that reactive tracer consumption in any parcel of test solution begins immediately after its injection into the aquifer, i.e., some reactive tracer is already being consumed during the finite-length injection phase. With this assumption, the extraction-phase breakthrough curve for a reactant is given by:

$$C_r^*(t^*) = \frac{C_{tr}^*(t^*)}{k\, T_{inj}}\left[e^{-kt^*} - e^{-k\left(T_{inj}+t^*\right)}\right] \tag{4.17}$$

where t^* is time elapsed since the end of the test solution injection, and T_{inj} is duration of the injection phase (Fig. 4.9a). Equation (4.17) can be rewritten as:

$$\ln\left(\frac{C_r^*(t^*)}{C_{tr}^*(t^*)}\right) = \ln\left[\frac{\left(1 - e^{-kT_{inj}}\right)}{k\, T_{inj}}\right] - k\, t^* \tag{4.18}$$

so that a plot of $\ln(C_r^*(t^*)/C_{tr}^*(t^*))$ versus t^* generates a straight line with a slope $-k$ and an intercept $\ln[(1-e^{-k\,T_{inj}})/k\,T_{inj}]$. Nonlinear least-squares regression may be used to fit Eq. 4.18 to experimental breakthrough data to obtain estimates of k. Subsequently, 95 % confidence intervals for k may be computed from the variance of k as described in Schroth et al. (1998). For more details on method development the reader is referred to Haggerty et al. (1998).

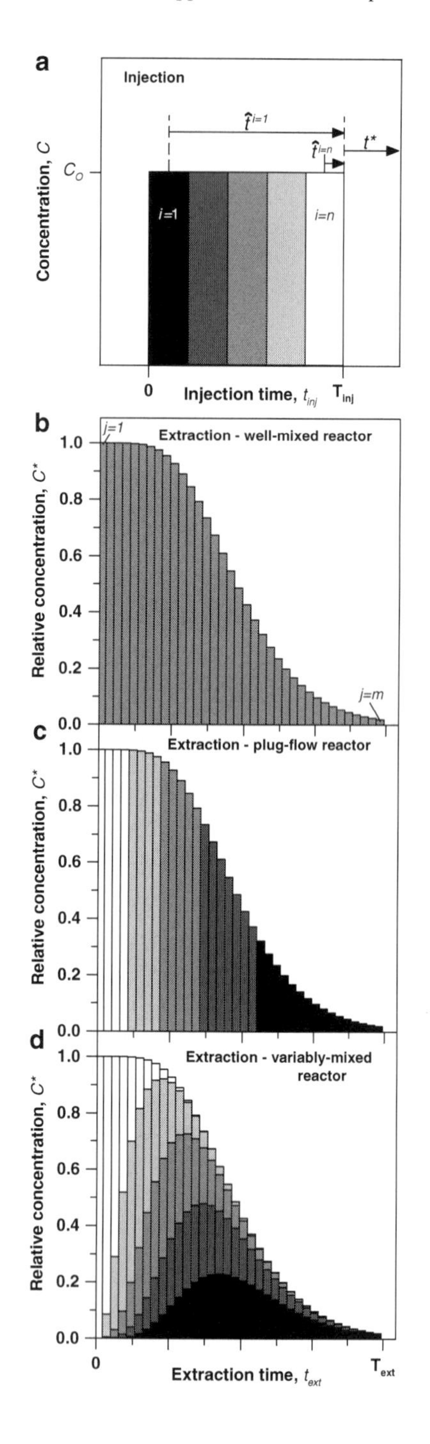

Fig. 4.9 Sketch of solute concentrations observed during a hypothetical push-pull test. The changing shading shows the changing composition of the injected test solution as it mixes with ambient groundwater: (**a**) the test solution is divided into individual parcels $i = 1$ to n, which are sequentially injected during the injection phase. (**b–d**) Composition of parcels $j = 1$ to m collected during the test's extraction phase for three different mixing models (Schroth and Istok 2004)

4.4.2 Plug-Flow Reactor Model

In contrast to the well-mixed reactor model, the plug-flow reactor model assumes that no mixing of injected test solution occurs in the portion of the aquifer investigated by the push-pull test. As a consequence of this assumption and because of the flow reversal that occurs during a push-pull test, the first parcel of test solution injected is the last to be extracted and the last parcel injected is the first to be extracted, i.e., each fluid parcel retains its identity during the test (the difference between the well-mixed and plug-flow reactor models can be seen by comparing Figs. 4.9b and c). As before, dilution of test solution with ambient ground water is separately accounted for using the nonreactive tracer data. This allows us to calculate the residence time for each parcel j, $t^{j}_{r,pf}$ using the portion of total tracer recovered at the time parcel j is extracted:

$$t^{j}_{r,pf} = t^{*j} + \frac{\int_{t_{ext}=0}^{t^{j}_{ext}} Q_{ext} C_{tr}(t)dt}{M_{tr}} T_{inj} \tag{4.19}$$

where Q_{ext} is the extraction pumping rate, t_{ext} is time since extraction began, and M_{tr} is total mass of tracer injected. A similar calculation was previously used to estimate Michaelis-Menten parameters from reaction product formation (Istok et al. 2001). Rate estimates are obtained by fitting

$$\ln\left(\frac{C^{*}_{r}(t^{*})}{C^{*}_{tr}(t^{*})}\right) = -k\, t_{r,pf} \tag{4.20}$$

to experimental data plotted as $\ln(C_r^{*}(t^{*})/C_{tr}^{*}(t^{*}))$ versus $t_{r,pf}$. Note that in this linear regression analysis the y-axis intercept of the fitted line must be forced through zero to satisfy $\ln(C_r^{*}(t^{*})/C_{tr}^{*}(t^{*})) = 0$ at $t_{r,pf} = 0$.

4.4.3 Variably-Mixed Reactor Model

Finally, we consider variable mixing of the injected test solution, which may occur during a push-pull test as a consequence of different distances traveled by individual parcels i and the variable velocity field encountered near the injection/extraction well (e.g., Gelhar and Collins 1971). In this case parcels injected early during the injection phase (e.g., i = 1, Fig. 4.9a) travel a greater distance from the injection/extraction well and exhibit a larger spread during extraction as compared to parcels injected late during the injection phase (Fig. 4.9d). Consequently, during extraction each parcel j is composed of different fractions of parcels i = 1 to n. When this composition is known, a weighted mean residence time ($t_{r,wm}$) can be computed for each parcel j using

$$t^{j}_{r,wm} = t^{*j} + \frac{\sum_{i=1}^{n}\left(C^{*\,i,j}_{tr}\,\hat{t}^{i}\right)}{\sum_{i=1}^{n} C^{*\,i,j}_{tr}} \tag{4.21}$$

where $\hat{t}^{i}$ is time elapsed from the midpoint of injection of parcel i to T_{inj} (Fig. 4.9a). Estimates of k can now be obtained from

$$\ln\left(\frac{C^{*}_{r}(t^{*})}{C^{*}_{tr}(t^{*})}\right) = -k\, t_{r,wm} \tag{4.22}$$

in a similar fashion as the completely mixed reactor model. Likewise, the y-axis intercept of the fitted linc must be forced through zero to satisfy $\ln(C_r^{*}(t^{*})/C_{tr}^{*}(t^{*})) = 0$ at $t_{r,wm} = 0$.

4.4.4 Examples

To illustrate these methods, a series of push-pull tests was conducted in a shallow, unconfined aquifer. Monitoring wells were constructed of 5.1 cm PVC casing and screen and were installed to a depth of 5 m. Water table depth prior to testing was 1.24 m. Test solutions were prepared from tap water and contained 100 mg/L Br^{-} as a nonreactive tracer (from KBr) and 25 mg/L NO_3^{-} (from $NaNO_3$) as the single reactive tracer to determine the in situ denitrification rate (defined here as the loss of NO_3^{-} from solution). Test solutions were prepared in 50 L plastic carboys. The test solution was mixed, and dissolved oxygen was removed, by vigorously bubbling compressed N_2 gas through the test solution until measured dissolved oxygen concentrations were reduced to <1 mg/L. The test solution was injected at ~1 L/min using a peristaltic pump; six samples of the test solution were collected during injection. An additional 10 L tap water "chaser" containing no added solutes was injected immediately after test solution to force all injected test solution into the formation. Compressed N_2 gas was used to mix the contents of the well casing during injection and extraction. Because the rate of denitrification was anticipated to be fairly large, extraction pumping began immediately after chaser injection was completed. During the extraction phase, the well was pumped at ~1 L/min and samples were collected after every 2 L had been extracted. All samples were analyzed for injected Br^{-} and NO_3^{-} by ion chromatography.

Breakthrough curves were prepared by plotting $C^{*} = C/C_o$ for Br^{-} and NO_3^{-}, where C is the measured concentration in an extraction sample, and C_o is the average concentration of the same tracer in the injected test solution, versus elapsed time since the end of test solution injection, t* (Fig. 4.10). Because the transport properties of Br^{-} and NO_3^{-} are expected to be similar in the absence of microbial activity, the observed differences in breakthrough curves for the two tracers is

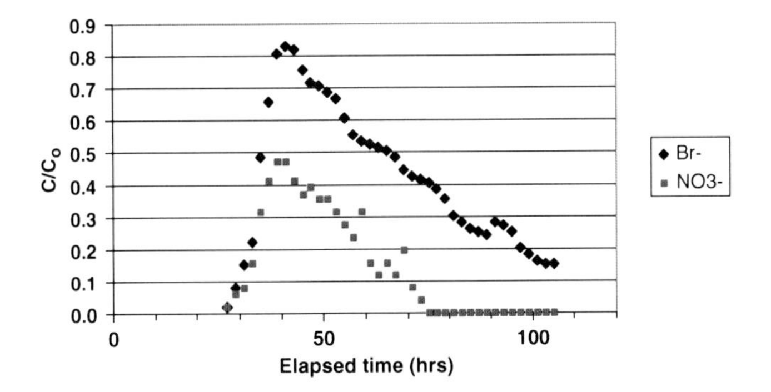

Fig. 4.10 Extraction phase breakthrough curves for push-pull test conducted in anaerobic, petroleum contaminated aquifer. Smaller relative concentrations of the injected reactive tracer NO_3^- relative to the coinjected nonreactive tracer Br^- are attributed to denitrifying activity of indigenous microorganisms

Fig. 4.11 Dilution-adjusted NO_3^- concentrations for data from Fig. 4.10 showing fitted regression line used to estimate zero-order reaction rate for NO_3^- reduction

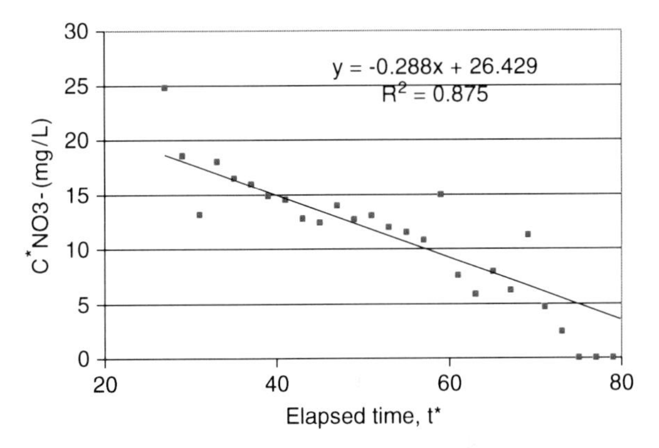

attributed to the activity of indigenous denitrifying bacteria. This interpretation was supported by the observed production of NO_2^-, and the observed accumulation of N_2O in other tests when acetylene was included in injected test solutions as an inhibitor of the enzyme system that reduces N_2O to N_2 (see Schroth et al. 1998).

The effects of dilution on the NO_3^- breakthrough curve can be removed by plotting dilution-adjusted NO_3^- concentrations:

$$C_{sol}^* = \frac{C_{sol}}{\left(\frac{C}{C_o}\right)_{tr}} \tag{4.23}$$

where C_{sol}^* is the dilution-adjusted NO_3^- concentration in an extraction phase sample, and $(C/C_o)_{tr}$ is the relative Br^- concentration in the same sample. These data can be used to observe the progress of the nitrate reduction reaction and estimate the apparent zero-order rate constant for nitrate consumption (Fig. 4.11). The fitted rate is 0.288 mg/L min.

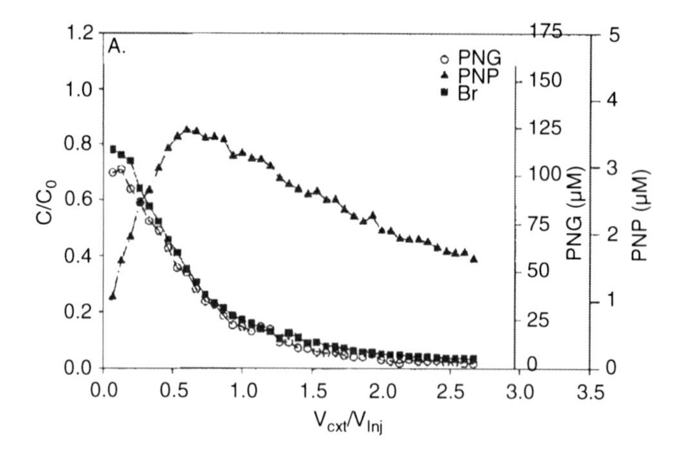

Fig. 4.12 Results of field push-pull test showing in situ PNP production from injected PNG (Istok et al. 2001)

Of course, push-pull tests may be used to estimate reaction rate parameters for more complicated reaction rate expressions than the simple first-order reaction used in this example (Eq. 4.15). As another example, Istok et al. (2001) used push-pull tests to determine Michaelis-Menton reaction kinetic parameters for the β-glucosidase enzyme system (which is related to total microbial community size). In these tests, β-glucosidase activity was quantified by injecting *p*-nitrophenyl-β-D-glucopyranoside (PNG) in a push-pull test and measuring the rate of in situ formation of PNP. The hydrolysis reaction conforms to Michaelis-Menton kinetics:

$$\frac{dC}{dt} = \frac{V_{max} C}{K_m + C} \tag{4.24}$$

where C is substrate concentration (PNG in this example), V_{max} is the maximum rate of substrate utilization, and K_m is the Michaelis constant, which is equal to the substrate concentration when $dC/dt = V_{max}/2$. An example breakthrough curve showing in situ PNP production is in Fig. 4.12 and the fitted Michaelis-Menten reaction rates are shown in Fig. 4.13.

4.5 Anaerobic Transformation of Chlorinated Solvents

Push-pull tests have been conducted to investigate microbial transformations of a wide variety of contaminants in the subsurface including chlorinated solvents, petroleum hydrocarbons, metals, and radionuclides. The basic idea behind these applications is simple; a test solution is prepared that contains a nonreactive tracer, one or more potentially reactive tracers (often the contaminant of interest), and other solutes as needed to investigate the targeted reactions (i.e., carbon or energy

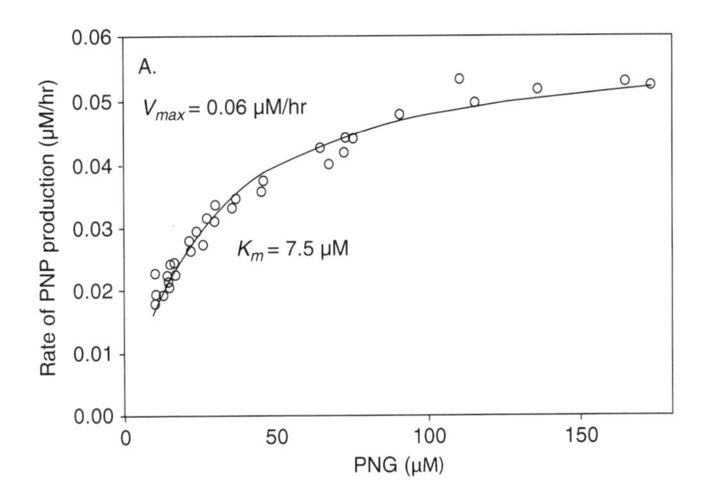

Fig. 4.13 Fitted Michaelis-Menten reaction rate parameters to PNP breakthrough curves in Fig. 4.12

substrates, growth limiting nutrients, etc.). If the contaminants of interest are present in the background porefluids, it may be necessary to use stable-isotope, radioactive-isotope, or chemically labeled-contaminants as reactive tracers when preparing the test solution so that contaminants and transformation products formed during the test can be distinguished from those already present in the formation.

In a series of field tests, trichlorofluoroethene (TCFE) was used to investigate microbial transformations of the common groundwater contaminant, trichloroethene (TCE). Laboratory and field research has shown that microbial transformation pathways for TCFE and TCE under anaerobic conditions are similar, consisting of a series of sequential reductive dechlorination reactions wherein one chlorine atom is replace by a hydrogen atom (Fig. 4.14). Because TCFE contains the fluorine label, it is possible to monitor microbial transformations of injected TCFE in the presence of background TCE and its transformation products using gas chromatography/mass spectroscopy (e.g., Hageman et al. 2001).

Hageman et al. (2001) conducted a series of push-pull tests using TCFE in a TCE-contaminated groundwater at a former chemical manufacturing plant. Background contaminants included TCE and tetrachloroethene (PCE), pesticides, petroleum hydrocarbons, and heavy metals. For this example, a push-pull test was conducted in a well that had previously been contaminated with TCE but prior to the test had no detectable TCE or its transformation products. The objective was to investigate microbial transformations of TCFE and TCE in the presence of added formate (to serve as an electron donor for reductive dechlorination). A 50 L test solution containing 100 mg/L Br^- (from KBr) and 2 mg/L formate (from sodium formate) was prepared from tap water in a plastic carboy and mixed with compressed Ar to remove dissolved oxygen prior to the start of the injection phase. A separate concentrated aqueous solution of TCFE and TCE was prepared in a collapsible metallized-film gas-sampling bag to prevent volatilization losses of

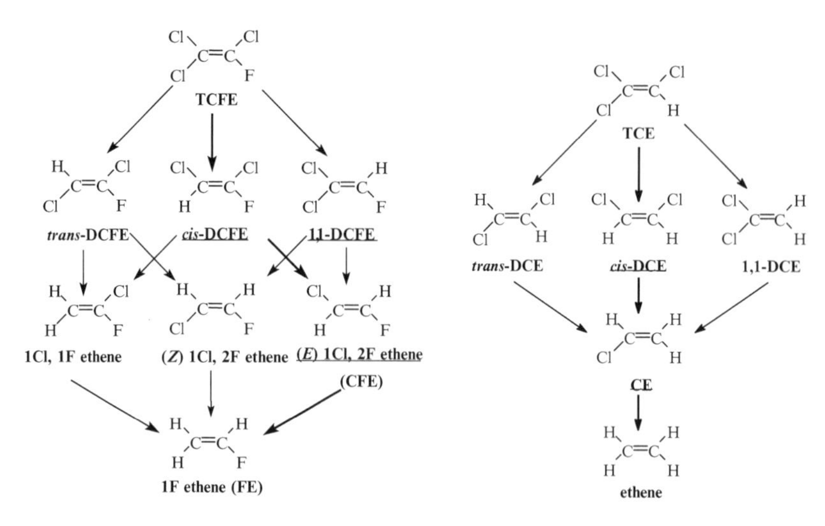

Fig. 4.14 Reductive dechlorination pathways for the contaminant TCE (*right*) and the reactive tracer TCFE (*left*). The predominant isomers and pathways are indicated by *underlines* and *heavy arrows*

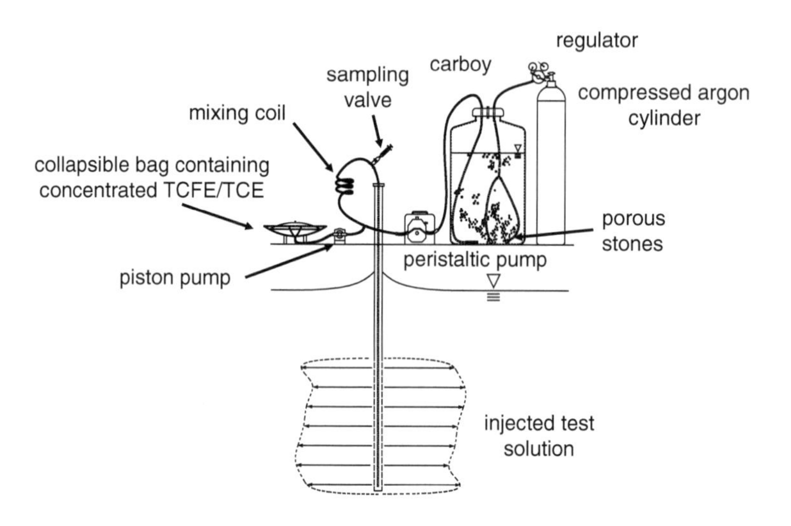

Fig. 4.15 Field equipment used to measure in situ rates of TCE and TCFE reductive dechlorination

TCFE and TCE during injection (Fig. 4.15). During the injection phase, the two solutions were combined using calibrated pumps to obtain the desired concentrations; a 10 m length of tubing was used as a "mixing coil" to allow sufficient time for turbulence to mix the two fluids before they entered the formation. Samples of the test solution were collected from the sampling valve during injection. The injection rate was 0.2 L/min. Because the reductive dechlorination of TCFE and TCE was

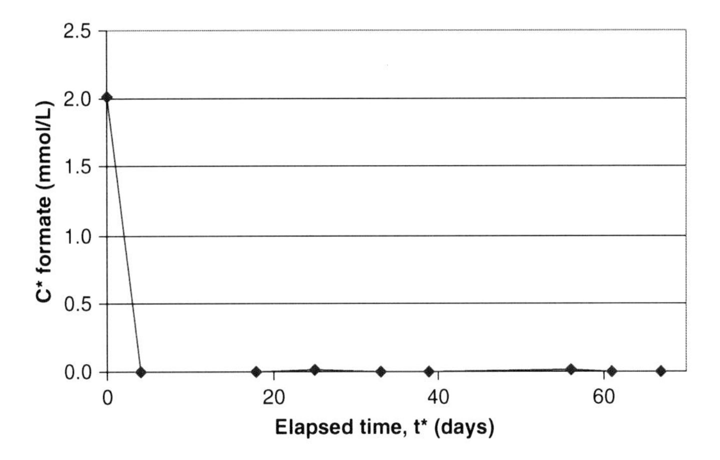

Fig. 4.16 Dilution-adjusted concentrations showing rapid utilization of injected formate

anticipated to be a relatively slow process, samples of the test solution/groundwater mixture were collected approximately once per week for ~10 weeks and analyzed for Br^- by ion chromatography and TCFE and TCE and their transformation products by gas chromatography/mass spectrometery (Hageman et al. 2001).

Measured concentrations of the reactive tracers were adjusted for dilution by dividing measured concentrations by a dilution factor. The dilution factor for nonsorbing reactive tracers is the relative concentration of the nonreactive and nonsorbing tracer (Br^- in this example) as described in Example 4. Since formate ($HCOO^-$) was not expected to sorb to these sediments due to its high water solubility and negative charge, measured formate concentrations were adjusted for dilution using relative concentrations of the co-injected bromide tracer:

$$C^*_{\text{formate}} = \frac{C_{\text{formate}}}{\left(\frac{C}{C_o}\right)_{Br}} \tag{4.25}$$

where C^*_{formate} is the dilution-adjusted formate concentration, C_{formate} is the measured formate concentration in a sample, C is the measured Br^- concentration in the same sample, and C_o is the average Br^- concentration in the injected test solution. Formate utilization was extremely rapid as shown in Fig. 4.16.

Since previous field testing indicated that TCFE, TCE, and their transformation products sorbed to aquifer sediments, Hageman et al. (2001) defined a dilution factor (DF) for each reactive tracer as the ratio of the summed concentrations of the tracer and its transformation products divided by the sum of tracer and transformation product concentrations in the injected test solution:

$$DF_{TCFE} = \frac{C_{TCFE} + C_{cis-DCFE} + C_{trans-DCFE} + C_{CFE} + C_{FE}}{(C_{TCFE} + C_{cis-DCFE} + C_{trans-DCFE} + C_{CFE} + C_{CFE})_o} \tag{4.26}$$

$$DF_{TCE} = \frac{C_{TCE} + C_{cis-DCE} + C_{trans-DCE} + C_{CE} + C_{Ethene}}{(C_{TCE} + C_{cis-DCE} + C_{trans-DCE} + C_{CFE} + C_{FE})_o} \tag{4.27}$$

$$C^*_{TCFE} = \frac{C_{TCFE}}{DF_{TCFE}}; C^*_{cis-DCFE} = \frac{C_{cis-DCFE}}{DF_{TCFE}};$$

$$C^*_{trans-DCFE} = \frac{C_{trans-DCFE}}{DF_{TCFE}}; C^*_{CFE} = \frac{C_{CFE}}{DF_{TCFE}} C^*_{CFE} = \frac{C_{CFE}}{DF_{TCFE}}$$

$$C^*_{TCE} = \frac{C_{TCE}}{DF_{TCE}}; C^*_{cis-DCE} = \frac{C_{cis-DCE}}{DF_{TCE}}; C^*_{trans-DCE} = \frac{C_{trans-DCE}}{DF_{TCE}};$$

$$C^*_{CE} = \frac{C_{CE}}{DF_{TCE}}; C^*_{CE} = \frac{C_{CE}}{DF_{TCE}}$$

Where for example, DF_{TCFE} is the dilution factor for TCFE and C^*_{TCFE} is the dilution-adjusted concentration for TCFE, etc. In this approach to dilution-adjustment, it is assumed that (a) all transformation products are identified and quantified in each sample and (b) the transport behaviors of TCFE and its transformation products are similar. The first assumption is considered valid because all known transformation products of TCFE and TCE (Fig. 4.14) were analyzed by GC/MS. The second assumption is supported by the results of additional field tests conducted at the site and by calculated retardation factors estimated for each compound using aquifer sediment properties (bulk density and organic matter content) and calculated organic matter partition coefficients for these compounds (Hageman et al. 2001).

This approach to dilution adjustment is called the *Forced Mass Balance Method* and was examined in detail by Hageman et al. (2003).

This test compared the transformation pathways and transformation rates of injected TCFE and TCE. TCFE was transformed to *cis*-DCFE, *trans*-DCFE, and (*E*)-1-chloro-2-fluoroethene (CFE) but fluoroethene (FE) was not detected (Fig. 4.17). Co-injected TCE was transformed to *cis*-DCE, *trans*-DCE, and 1,1-DCE (Fig. 4.18). CE was detected in the samples collected on days 56 and 67; however, its concentration was below the quantitation limit. Ethene was not detected.

Rates of TCFE and TCE transformation can be computed directly using the dilution-adjusted concentrations as shown in Figs. 4.17 and 4.18.

At many sites, microbial transformation of TCE results in the accumulation of vinyl chloride (VC), a known carcinogen and neurotoxin. Although qualitative evidence for the transformation of VC to ethene can be obtained by a number of field methods, quantitative tools are needed to determine the in-situ rates of VC transformation to ethene in contaminated groundwater. Ennis et al. (2005) investigated reductive dechlorination of chlorofluoroethene (CFE) as a surrogate reactive tracer for VC.

Commercial CFE contains a mixture of Z and E optical isomers. A concentrated aqueous solution containing Z-/E-CFE as reactive tracers and Br^- as a nonreactive

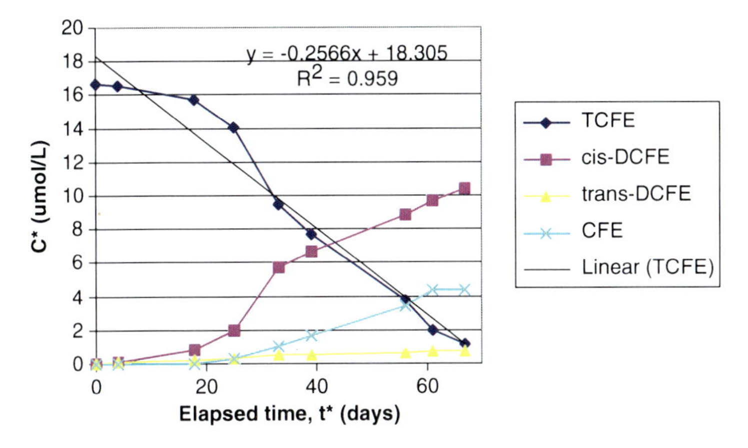

Fig. 4.17 Dilution-adjusted concentrations for injected TCFE and its transformation products *cis*-DCFE, *trans*-DCFE, and CFE formed in situ

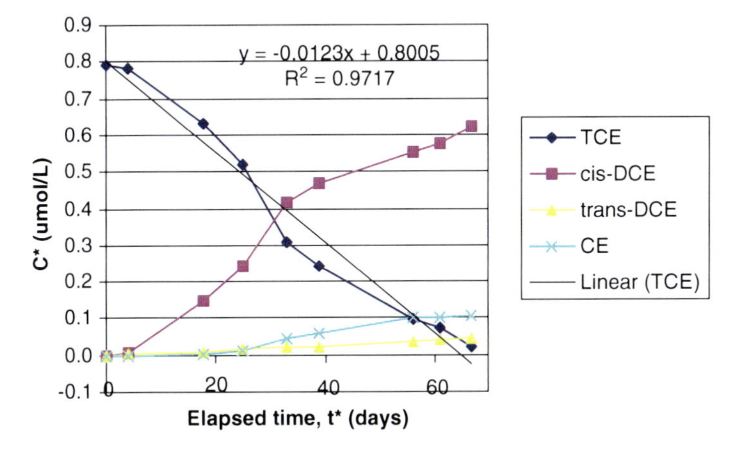

Fig. 4.18 Dilution-adjusted concentrations for injected TCE and its transformation products *cis*-DCE, *trans*-DCE, and CE formed in situ

tracer in distilled water was prepared in a metallized-film collapsible bag. A separate 50 L plastic carboy was filled with tap water. Calibrated peristaltic and piston pumps were used to combine the tap water from the carboy and the concentrated tracer solution from the collapsible bag during injection and obtain the targeted tracer concentrations (100–300 μM for CFE and ~1 mM for Br^-). ~52 L of the combined test solution were injected at a rate of ~0.2 L/min. Samples of the test solution/groundwater mixture were collected from the test well periodically for ~60 days. Samples were analyzed for injected CFE and Br^- and for FE formed in situ; analytical methods are described in Elliot et al. (2004).

Dilution-adjusted concentrations for CFE and FE were computed using the Forced Mass Balance method of Hageman et al. (2003). Results for a test conducted in well

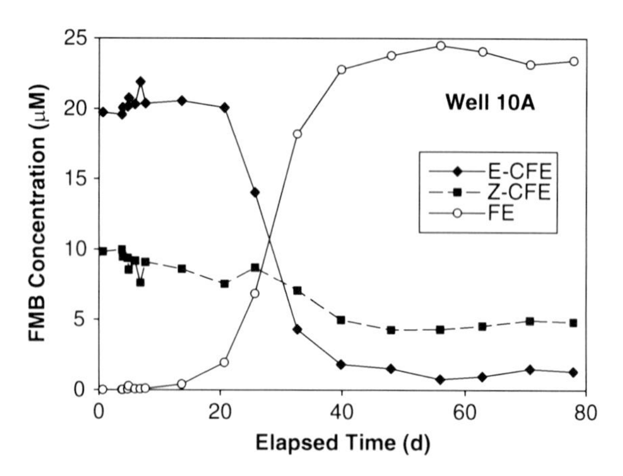

Fig. 4.19 Dilution-adjusted (forced mass balance or FMB) concentrations for push-pull test showing in situ reductive dechlorination of injected chlorofluoroethene (CFE) to fluoroethene (FE) (Elliot et al. 2004).

10A show the transformation of injected CFE to FE (Fig. 4.19). The maximum rates of E-CFE and Z-CFE disappearance (computed by regression using data collected between days 20 and 40) were 0.33 ± 0.02 and 0.08 ± 0.01 μM/day, respectively; the corresponding rate of FE formation was 0.40 ± 0.04 μM/day. The ratio E-CFE/Z-CFE decreased from 2.0 to 0.6 over the 80 day test period showing preferential transformation of the E-CFE isomer.

4.6 Anaerobic Transformations of Petroleum Hydrocarbons

Groundwater contamination by petroleum hydrocarbons, including benzene, toluene, ethylbenzene, and xylene isomers (BTEX) is widespread. Since many sites contaminated with these compounds are anaerobic, BTEX degradation under oxygen-limited conditions is of interest. Anaerobic BTEX degradation has been shown to occur under denitrifying, sulfate-reducing, iron-reducing, manganese-reducing, and methanogenic conditions. Benzylsuccinic acid (BSA) and methyl-BSA have been identified as products of the anaerobic metabolism of toluene and xylenes, respectively. The first intermediate of anaerobic toluene mineralization is BSA and methyl-BSA can be formed as either an intermediate of anaerobic xylene mineralization or a dead-end product of anaerobic xylene cometabolism, depending on the bacterial culture involved.

Reusseur et al. (2002) conducted a series of push-pull tests to detect and quantify the formation of deuterated benzylsuccinic acid (BSA-d_8) and *o*-methyl benzylsuccinic acid (*o*-methyl BSA-d_{10}) resulting from the injection of deuterated toluene-d_8 and *o*-xylene-d_{10}, respectively. The expected anaerobic transformation pathways are shown in Fig. 4.20. One set of tests was conducted in unconfined aquifer consisting of ~5 m of medium dense to fine grained sand and silty sand overlaying clayey silt with sand interbedded with silty clays and clays. The water table was ~2–3 m below land surface and estimated groundwater velocity was

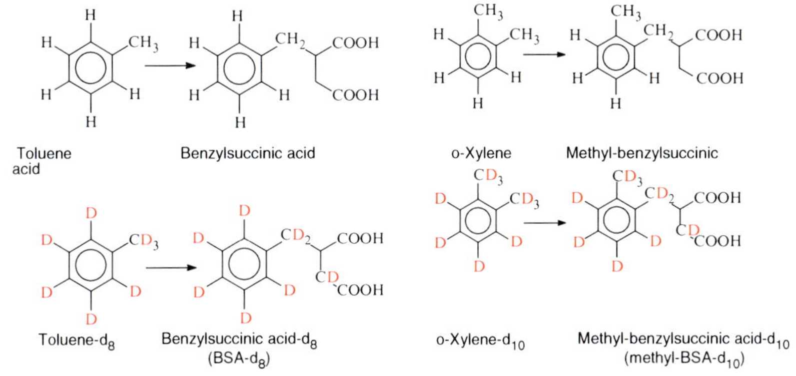

Toluene acid
Benzylsuccinic acid
o-Xylene
Methyl-benzylsuccinic
Toluene-d_8
Benzylsuccinic acid-d_8 (BSA-d_8)
o-Xylene-d_{10}
Methyl-benzylsuccinic acid-d_{10} (methyl-BSA-d_{10})

Fig. 4.20 Anaerobic transformation pathways for toluene and o-xylene (*upper*) and their deuterium labeled analogs (*lower*)

100 m/year. Total BTEX concentrations in wells used at the site were 71.5 and 193 μM. Tests were conducted in 5 cm inner diameter PVC monitoring wells with 3-m screened intervals starting at 1.6 m below land surface.

Test solutions consisted of tap water containing Br^- as a nonreactive, nonsorbing tracer (from KBr), nitrate (from $NaNO_3$) as a reactive, nonsorbing tracer, and toluene-d_8 and *o*-xylene-d_{10} as reactive, sorbing tracers. Sulfate (0.1 mM) was also present in the tap water used to prepare test solutions. Field equipment was identical to that used in the tests in Example 5. Test solutions were prepared and mixed in plastic tanks and then stripped of dissolved oxygen by vigorous bubbling with compressed Ar gas prior to injection. Toluene-d_8 and *o*-xylene-d_{10} were then introduced into the test solution during injection by a second pump that delivered a concentrated aqueous solution that had been prepared in a collapsible metallized bag. 250 L of the test solution were injected at a rate of 0.5–2 L/min. Samples were collected during the injection phase and daily-to-weekly for 30 days after injection. All samples were analyzed for bromide, nitrate, and sulfate by ion chromatography and for toluene-d_8, and *o*-xylene-d_{10} by GC/MS. Analytical details are in Reusseur et al. (2002).

Unambiguous evidence for in situ anaerobic transformation of injected toluene-d_8 and *o*-xylene-d_{10} was obtained from the observed production of their deuterated transformation products, BSA-d_8 and *o*-methyl-BSA-d_{10}, respectively, in well CR15 (Fig. 4.21). BSA-d_8 and *o*-methyl-BSA-d_{10} were detected beginning on day 5 with measured concentrations of 4.7 and 1.5 nM, respectively. The concentrations increased until day 8 with measured BSA-d_8 and *o*-methyl-BSA-d_{10} concentrations of 5.6 nM and 1.7 nM, respectively, which then decreased to below detection on day 9. Dilution-adjusted concentrations of BSA-d_8 and *o*-methyl-BSA-d_{10} were calculated by dividing their measured concentrations by the corresponding C/Co for Br. For example, on day 7 the measured BSA-d_8 concentration of 5.2 nM was divided by the C/Co for Br of 0.354 (which indicates 64.6 % dilution of the test solution) to obtain a dilution-adjusted BSA-d_8 concentration of 14.8 nM. Initial zero-order

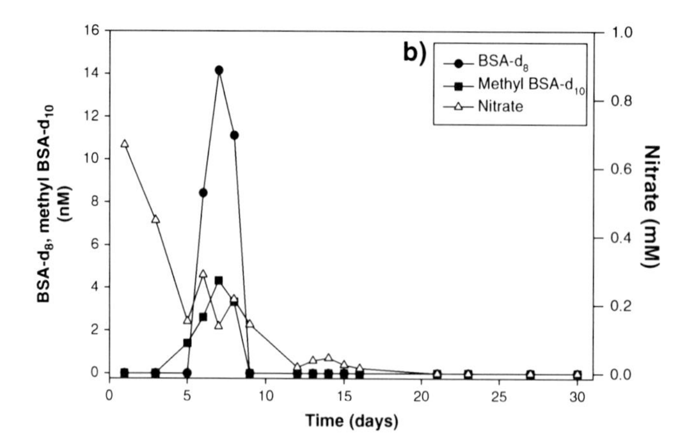

Fig. 4.21 Dilution-adjusted concentration profiles showing consumption of injected nitrate and formation of BSA-d_8 and *o*-methyl-BSA-d_{10} from injected toluene-d_8, and *o*-xylene-d_{10} (Reusseur et al. 2002)

formation rates for BSA-d_8 (7.4 nM/day) and *o*-methyl-BSA-d_{10} (1.0 nM/day) were estimated from dilution-adjusted concentrations determined on days 5–7.

4.7 Aerobic Cometabolism of Chlorinated Solvents

Kim et al. (2004) investigated aerobic cometabolism of chlorinated solvents in an aquifer contaminated with TCE and cis-DCE. By conducting a series of push-pull tests they were able to stimulate indigenous microbial activity including propane and oxygen utilization, the transformation of ethylene and propylene to their corresponding epoxides, and the cometabolic transformation of cis-DCE. By injecting acetylene they were able to inhibit propane utilization, ethylene and propylene transformation, and cis-DCE transformation confirming that monooxygenase was mainly responsible for the observed microbial activity.

Tests were performed in monitoring wells at a site contaminated with cis-DCE (20–40 μg/L) and TCE (200–400 μg/L). The aquifer consists primarily of alluvial deposits, and is unconfined with a water table depth ranging from 30 to 32 m below ground surface; groundwater at the site is aerobic with ~6 mg/L dissolved oxygen. Monitoring wells used in the tests were constructed of 5.1 cm polyvinyl chloride casing with a 2.9 m long screen.

Field equipment consisted of compressed and liquefied gases, gas flow meters, two carboys, a collapsible bag, a peristaltic pump to inject the test solution into the well, and a submersible pump to extract groundwater from the same well. Injected test solutions were prepared from three separately-prepared solutions: (1) KBr and $NaNO_3$ were added to 500 L of site groundwater in a plastic tank and bubbled with oxygen to obtain known concentrations of Br^- (nonreactive tracer), NO_3^- (nitrogen

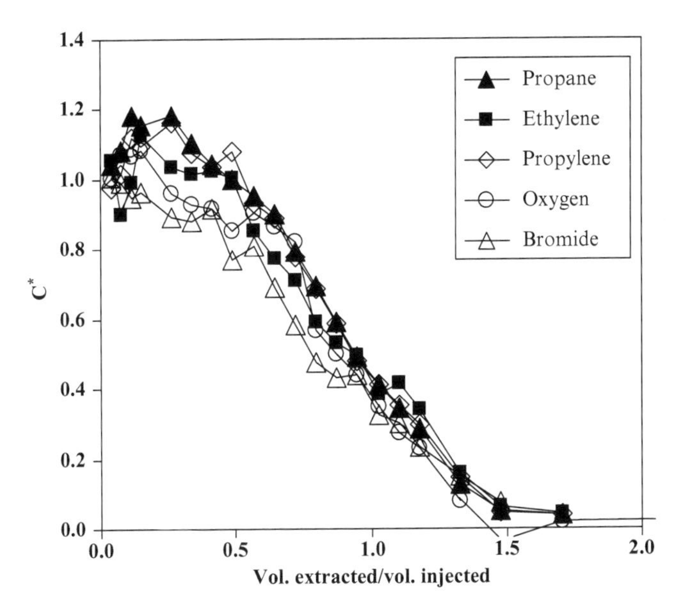

Fig. 4.22 Pull phase normalized concentrations during a Transport Test showing conservative (nonsorbing, nonreacting) transport of reactive tracers prior to biostimulation (Kim et al. 2004)

source), and dissolved oxygen (electron acceptor); (2) Various combinations of gases were bubbled through 50 L of site groundwater in a separate plastic carboy to obtain known concentrations of propane, ethylene, and/or propylene to probe for microbial activity; and (3) Known amounts of acetylene gas were added directly to 5 L of site groundwater in a collapsible bag to obtain known concentrations of dissolved acetylene. The three solutions were combined using calibrated peristaltic pumps and piton pumps as needed to obtain the desired tracer concentrations, which were verified by collecting samples during the injection phase and subsequently analyzing them by ion and gas chromatography using methods in Kim et al. (2004).

Short-duration "transport" tests were conducted to compare the relative mobility of Br^-, NO_3^-, and dissolved oxygen, propane, propylene, and ethylene. In these tests, 260 L of test solution containing all of these tracers (prepared as described above) were injected at 2 L/min. After a 16 h rest phase with no pumping, the test solution/ground water mixture was extracted from the well at a rate of 2.5 L/min. Extraction phase breakthrough curves for all injected solutes were similar and essentially all the injected solute mass was recovered during transport tests indicating conservative transport (no sorption and no reaction) of all injected tracers prior to biostimulation (Fig. 4.22). These results are important because they mean that measured concentrations of the reactive tracers can be adjusted for dilution using measured Br^- concentrations (Haggerty et al. 1998).

Microbial activity was then stimulated by five sequential additions of groundwater containing NO_3^- and dissolved propane and oxygen. Increased consumption of injected propane and oxygen were observed following each addition (Kim et al. 2004). Then a series of "Activity Tests" were conducted to quantify rates of propane utilization, ethylene and propylene transformation, and c-DCE and TCE transformation. Test solutions were prepared and injected as described for the "transport tests" described above but since microbial activity had been stimulated by sequential propane, oxygen, and nitrate additions, reactive tracers were rapidly consumed during the Activity Tests. "Acetylene Blocking" tests were conducted similarly to the "Activity Tests" except that dissolved acetylene (10 mg/L) was included in injected test solutions.

Mass balance calculations were performed by integrating measured solute concentrations and injection and extraction volumes. Concentration profiles for all reactive tracers were computed by plotting normalized concentrations, C*, versus time where

$$C^{*} = [(C - C_{BG})/(C_{o} - C_{BG})] \quad (4.28)$$

and C is the measured tracer concentration in a sample collected after injection, C_o is the average concentration of the same tracer in the injected test solution, and C_{BG} is the background (pre-injection) concentration of the same tracer (usually only important for oxygen) in the ambient groundwater.

Example results show the consumption of injected oxygen, propane, nitrate, and ethylene and the in situ production of ethylene oxide (Fig. 4.23). In the presence of added acetylene, reactive tracer consumption was essentially completely inhibited, and very little ethylene oxide was produced (Fig. 4.24). The strong inhibition by acetylene indicates that a propane monooxygenase enzyme is likely responsible for the observed propane utilization and cometabolism of ethylene.

Reactive tracer concentrations for a propane Activity Test (Fig. 4.23a) and Acetylene Blocking test (Fig. 4.23b) are plotted as 1-C*, that is, $1 - [(C - C_{BG})/(C_o - C_{BG})]$. This method of plotting was used because, unlike the other tracers, cis-DCE and TCE concentrations were lower in the injected test solution than in the background groundwater as a result of the sparging of site groundwater with oxygen and the other gases prior to injection. For an injected nonreactive tracer (Br^- in this test) values of 1-C* should be 0 at the end of the injection phase and increase to 1 as the injected test solution is diluted by ambient groundwater. Values for 1-C* for injected reactive tracers (oxygen, propane, or ethylene) should be greater than zero at the end of the injection phase and increase to 1 as the injected test solution is diluted by ambient groundwater. For reactive tracers with high concentrations in the ambient groundwater (cis-DCE or TCE) than the injected test solution, values of 1-C* should be negative at the end of the injection phase and then gradually increasing with time. During a propane Activity Test (Fig. 4.23a), values of 1-C* for propane and ethylene values were greater than 0 during the early portion of the extraction phase, and increased to unity as extraction continued, suggesting significant utilization of propane and transformation of ethylene occurred during the rest phase. However, values of 1-C* for cis-DCE were smaller than those for Br^-, indicating that cis-DCE was cometabolically transformed during the test, while TCE values were essentially identical to those of bromide suggesting that no detectable TCE transformation occurred. During an Acetylene

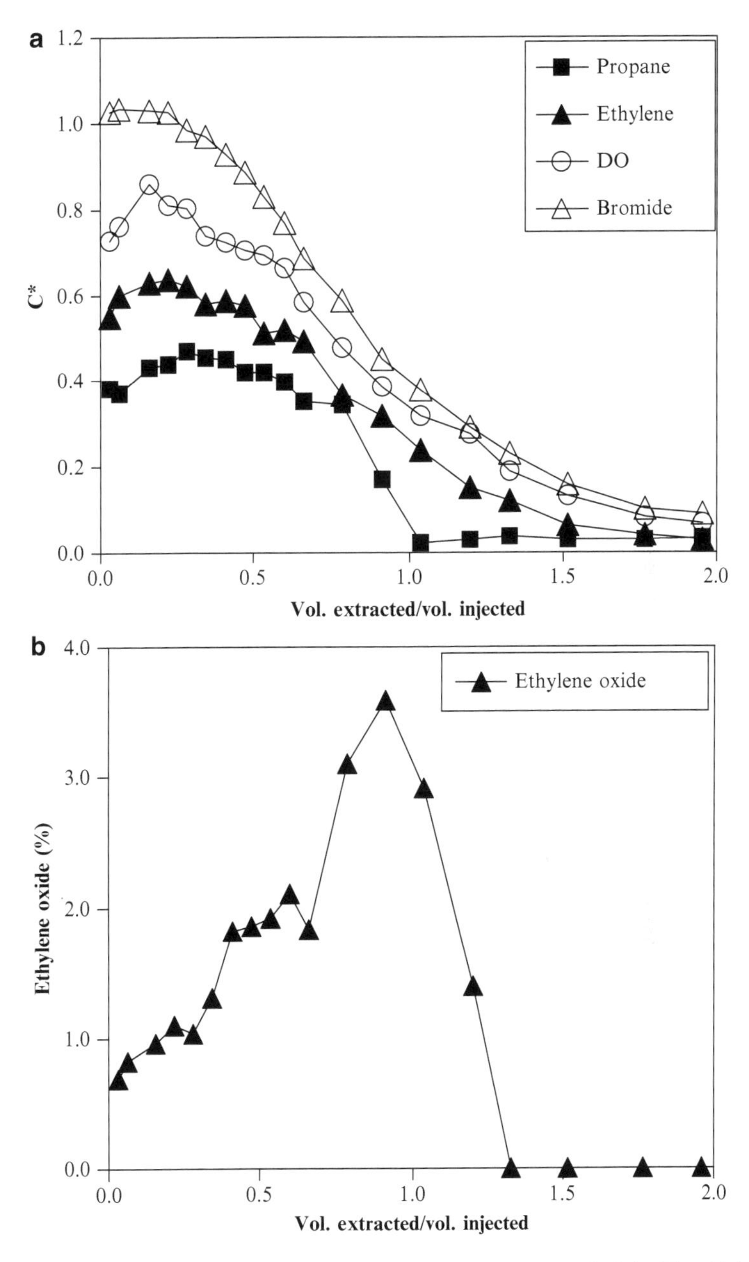

Fig. 4.23 Extraction phase breakthrough curves from during a propane Activity Test (**a**) injected tracers (**b**) ethylene oxide formed in situ (ethylene oxide concentrations are expressed as a percentage of the average ethylene concentration in injected test solution) (Kim et al. 2004)

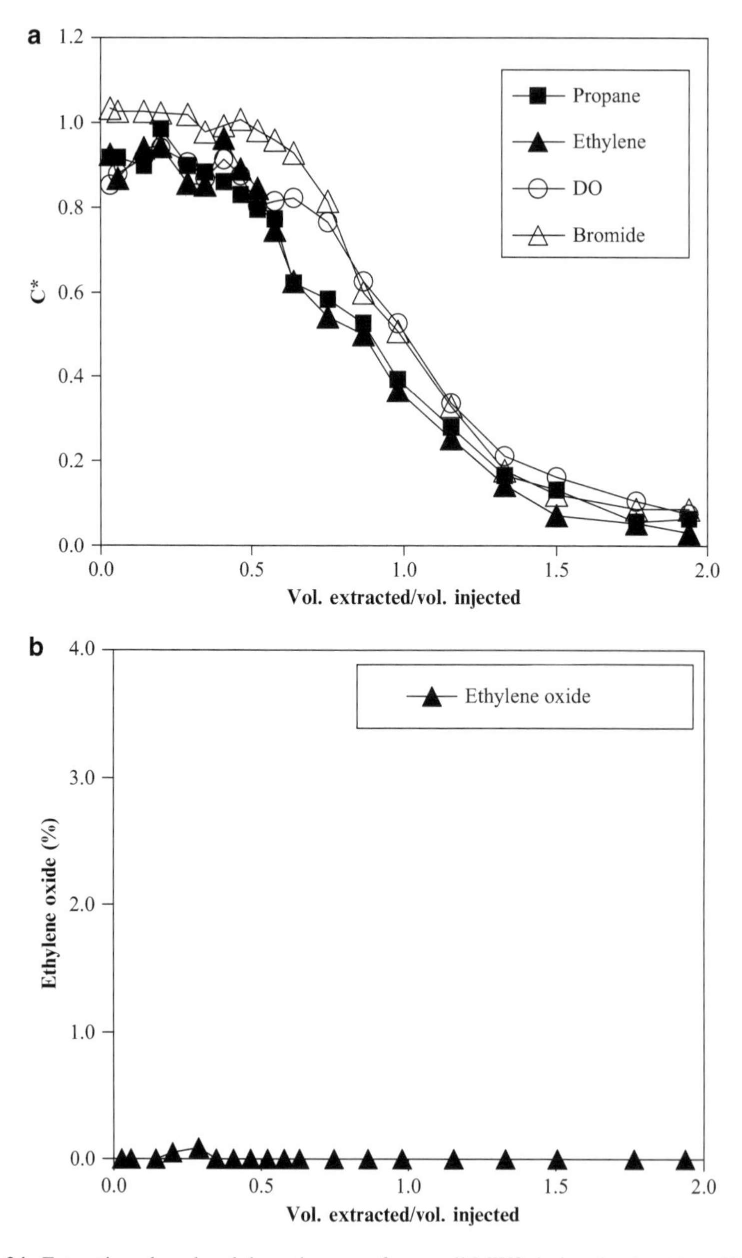

Fig. 4.24 Extraction phase breakthrough curves from well MW3 during the Acetylene Blocking Test (**a**) injected tracers (**b**) ethylene oxide formed in situ (ethylene oxide concentrations are expressed as a percentage of the average ethylene concentration in injected test solution) (Kim et al. 2004)

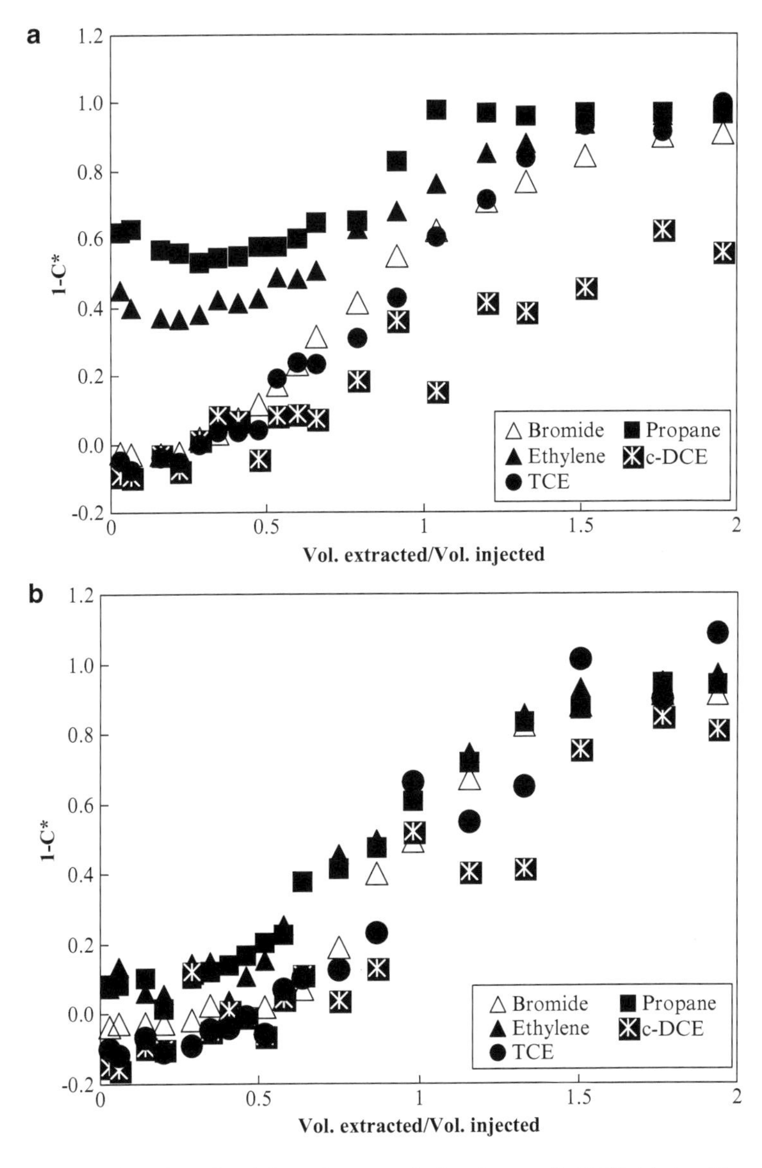

Fig. 4.25 Extraction phase breakthrough curves during from well MW3 (**a**) the 4th propane Activity Test, and (**b**) Acetylene Blocking Test. C_0 values for propane, ethylene, c-DCE, and TCE were ~1.4 mg/L, 2.1 mg/L, 5.0 μg/L and 40 μg/L, respectively. C_{BG} values for bromide, propane and ethylene were 0, while C_{BG} values for c-DCE and TCE were 15 and 160 μg/L, respectively

Blocking Test in contrast (Fig. 4.24b), values of 1-C* for all tracers were similar indicating that transformation of all reactive tracers including cis-DCE were inhibited by added acetylene (Fig. 4.25).

4.8 Anaerobic Transformation of Radionuclides

Contamination of the subsurface by U and ^{99}Tc (Tc) is common at certain industrial facilities associated with the nuclear fuel cycle or weapons production. In oxidizing environments, U occurs as U(VI), which forms highly soluble and mobile complexes with carbonate at pH >5 and Tc occurs as Tc(VII) in the form of the highly soluble and mobile pertechnetate anion (TcO_4^-). In reducing environments, U and Tc occur as U(IV) and Tc(IV), respectively, which have a much lower solubility and mobility than their oxidized forms. For this reason, bioimmobilization, the addition of nutrients to the subsurface to stimulate indigenous microorganisms to reduce U(VI) and Tc(VII) and precipitate U(IV) and Tc(IV) solid phases has been proposed as a strategy for reducing U and Tc concentrations in groundwater.

The potential to stimulate an indigenous microbial community to reduce U(VI) and Tc(VII) was evaluated in a shallow unconfined aquifer by Istok et al. (2004). Field tests were conducted in monitoring wells installed in the shallow unconfined aquifer formed in an unconsolidated silty-clayey saprolite. Wells were installed by direct-push methods to a depth of ~7 m and were constructed of 3 cm PVC with a 1.5 m screened interval at the bottom of the well.

Test solutions were prepared from site groundwater in plastic drums, amended with 1 mM Br^- as a nonreactive tracer, 40 mM ethanol as an electron donor, and contained 1 mM NO_3^-, 5 μM U(VI) and 400 pM Tc(VII). Approximately 200 L of test solution were injected for each test using a siphon over a period of 0.5–2 days. Five samples of the test solution were collected during injection. Following injection, groundwater samples were periodically collected from the same well for up to 400 days.

Microbial activity was not detected in "Control Tests" conducted without added electron donor. For example, in one test, relative concentrations (C/C_o, where C is the measured concentration in a sample and C_o is the average concentration of the same component in the injected test solution) of injected NO_3^-, U(VI), and Tc(VII) decreased similarly to injected Br^- (Fig. 4.26). This indicates that changes in NO_3^- and radionuclide concentrations were largely due to dilution of the test solution as it gradually drifted away from the injection well and that intrinsic rates of NO_3^-, U(VI) and Tc(VII) reduction (i.e., supported by endogenous electron donors) are not detectable and are likely very small. The results also confirm that, under the conditions of these tests (with added bicarbonate), sorption of injected NO_3^-, U(VI), and Tc(VII) to aquifer sediments at this site is negligible.

In tests conducted with added ethanol, however, injected NO_3^- was completely removed within ~180 h following injection and only trace levels of NO_2^- were detected (Fig. 4.27). Relative concentrations of U(VI) decreased continuously from $C/Co = 1$ to 0.03 by the end of the test (~ 400 h) with no apparent remobilization.

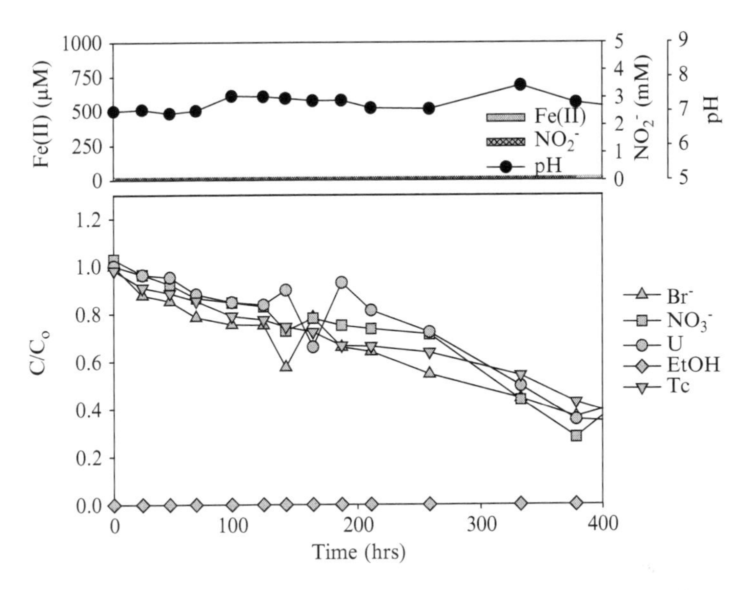

Fig. 4.26 Concentration profiles for "Control Test" showing dilution of injected test solution components and no NO_2^- or Fe(II) production in the absence of added ethanol (Istok et al. 2004)

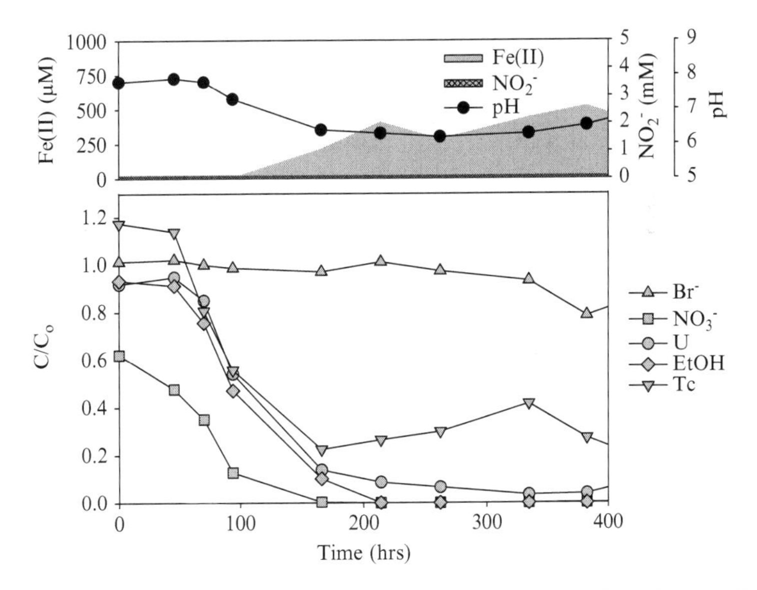

Fig. 4.27 Concentration profiles for test conducted with ethanol showing ethanol utilization, NO_3^-, Tc(VII), and U(VI) reduction and Fe(II) production (Istok et al. 2004)

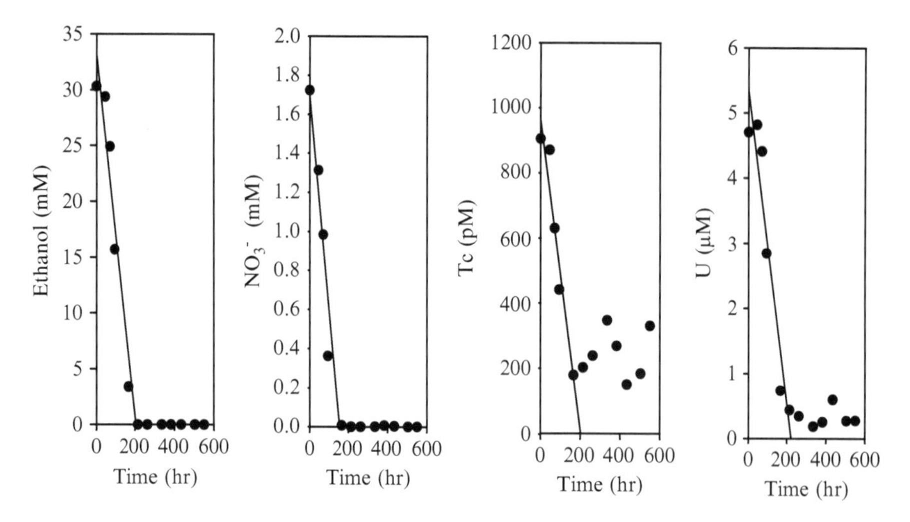

Fig. 4.28 Dilution-adjusted concentrations for push-pull test conducted with added ethanol showing regressions used to estimate rates of ethanol utilization and NO_3^-, Tc(VII), and U(VI) reduction (Istok et al. 2004)

Rates of donor utilization, NO_3^- reduction, and Tc(VII) reduction were computed from dilution-adjusted concentration profiles (Fig. 4.28) (Istok et al. 2004). For example, the rate of U(VI) reduction was 0.024 μM/h.

4.9 Quantification of Methane Oxidation in Soils Using Gas Push-Pull Tests

Methane oxidizing microorganisms (methanotrophs) are ubiquitous in soils and are potentially the largest biological sink for atmospheric methane. Urmann et al. (2005) conducted a series of gas phase push-pull tests to quantify methanotrophic activity in the vadose zone above a petroleum-contaminated aquifer. In these tests, a gas mixture containing the reactive tracer gases methane and oxygen and the nonreactive tracer gases argon and neon were injected into the vadose zone using a shallow well and then extracted from the same location.

To inject/extract gas mixtures under controlled conditions, Urmann et al. (2005) constructed a gas flow controller, which allows controlled injection and extraction of gases into and out of the soil vadose zone, gas sampling during both the injection and extraction phases of the test, as well as monitoring of gas pressure (Fig. 4.29).

In a typical test, 30 L of the gas mixture (N_2 containing ~3.2 vol. % Ne and 20 vol. % Ar and 3.3 vol. % CH_4 and 16.5 vol. % O_2) was injected at a flow rate of

Fig. 4.29 Field setup used in gas phase push-pull tests by Urmann et al. (2005)

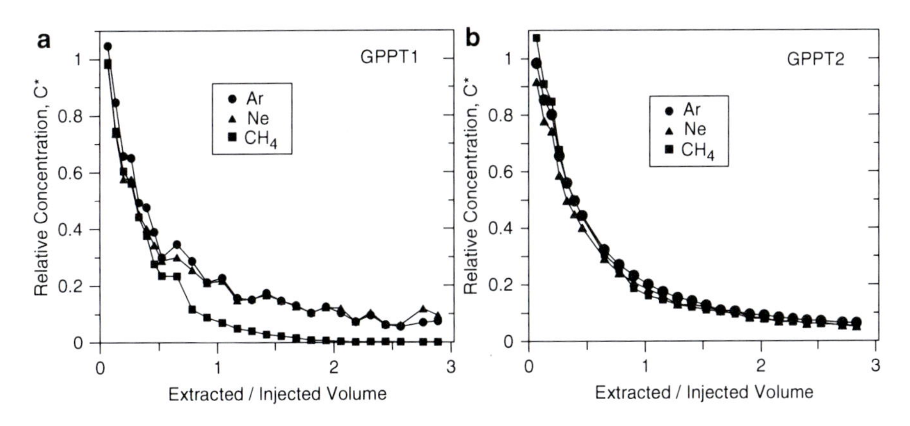

Fig. 4.30 Breakthrough curves for Ar, Ne, and CH_4 obtained during the extraction phase of two gas phase push-pull tests by Urmann et al. (2005). Acetylene was coinjected during GPPT2 to inhibit methane oxidation

~0.6 L min^{-1}. Extraction commenced immediately after injection, and a total of 90 L of gas mixture/ambient soil gas were extracted at a flow rate of 0.52 L min^{-1}. Total test duration was ~4.0 h. In some tests, ~2 vol. % C_2H_2 was included in the injected gas mixture to inhibit methanotrophic activity.

Results for two tests are presented here; GPTT1 (without acetylene) and GPTT2 (with coinjected acetylene). Relative concentrations, C* (concentration measured during extraction divided by the injection concentration) of Ne, Ar, and CH_4 declined during the extraction phase of both gas-push-pull tests as a consequence of dilution of the injected gas mixture with ambient soil gas (Fig. 4.30).

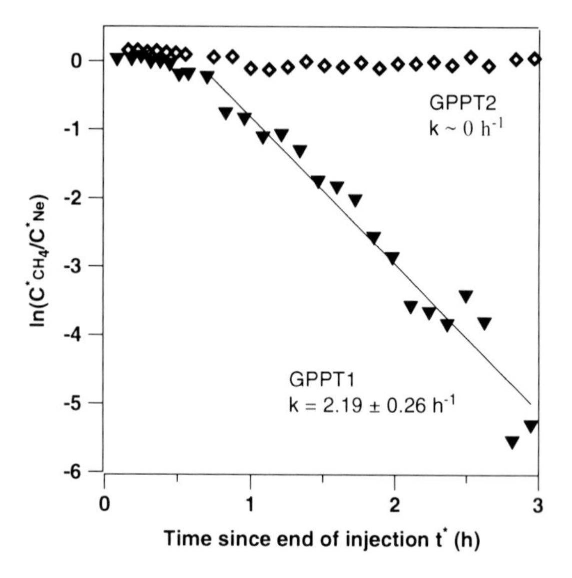

Fig. 4.31 Determination of first-order rate coefficients, k for methane oxidation for the two gas phase push-pull tests in Fig. 4.30 using the method of Haggerty et al. (1998) (Urmann et al. 2005)

Concentrations of injected CH_4 were substantially smaller than Ne or Ar concentrations during GPPT1 indicating microbial methane oxidation, which is confirmed by the results of GPPT2 with added acetylene where the relative concentrations of all injected gases were essentially identical. The estimated first-order rate coefficients for methane oxidation were 2.19 h^{-1} for GPPT1 and ~0 h^{-1} for GPTT2 (Fig. 4.31).

References

Beauheim RL (1987) Interpretations of Single-Well Hydraulic Tests Conducted At and Near the Waste Isolation Pilot Plant (WIPP) Site, 1983–1987, SAND87-0039, Albuquerque, NM: Sandia National Laboratories

Borden RC, Lee MD, Thomas JM, Bedient PB, Ward CH (1989) In situ measurement and numerical simulation of oxygen limited biotransformation. Ground Water Monit Rev IX(1):83–90

Burbery L, Cassianib G, Andreottic G, Ricchiutod T, Semplea KT (2004) Single-well reactive tracer test and stable isotope analysis for determination of microbial activity in a fast hydrocarbon-contaminated aquifer. Environ Pollut 129:321–330

Davis BM, Istok JD, Semprini L (2002) Push-pull partitioning tracer tests using radon-222 to quantify non-aqueous phase liquid contamination. J Contam Hydrol 58:129–146

Davis BM, Istok JD, Semprini L (2003) Static and push-pull methods using radon-222 to characterize dense nonaqueous phase liquid saturations. Ground Water 41(4):470–481

Davis BD, Istok JD, Semprini L (2005) Numerical simulations of radon as an in situ partitioning tracer for quantifying NAPL contamination using push–pull tests. J Contam Hydrol 78:87–103

Drever JI, McKee CR (1980) The push-pull test a method of evaluating formation adsorption parameters for predicting the environmental effects on in-situ coal gasification and uranium recovery. In Situ 4(3):181–206

Ennis E, Reed R, Dolan M, Semprini L, Istok JD, Field JA (2005) Reductive dechlorination of the vinyl chloride surrogate chlorofluoroethene in TCE-contaminated groundwater. Environ Sci Technol 39:6777–6785

Field JA, Istok JD, Schroth MH, Sawyer TE, Humphrey MD (1999) Laboratory investigation of surfactant-enhanced trichloroethene solubilization using single-well "push-pull" tests. Ground Water 37(4):581–588

Field JA, Sawyer TE, Schroth MH, Humphrey MD, Istok JD (2000) Effect of cation exchange on surfactant-enhanced solubilization of trichloroethene. J Contam Hydrol 46:131–149

Field JA, Reed R, Istok JD, Semprini L, Bennett P, Buscheck TE (2005) Trichlorofluoroethene: a reactive tracer for evaluating reductive dechlorination in large-diameter permeable columns. Ground Water Monit Remed 25(2):68–77

Gelhar LW, Collins MA (1971) General analysis of longitudinal dispersion in nonuniform flow. Water Resour 7(6):1511–1521

Gomez K, Gonzalez-Gil G, Schroth MH, Zeyer J (2008) Transport of methane and noble gases during gas push-pull tests in variably-saturated porous media. Environ Sci Technol 42(7):2515–2521

Gonzalez-Gil G, Urmann K, Gomez K, Schroth MH, Zeyer J (2006) In situ quantification of methane oxidation in soils using gas push-pull tests. Int Cong Ser 1293:42–45

Gu B, Yan H, Zhou P, Watson DB, Park M, Istok JD (2005) Natural humics impact uranium bioreduction and oxidation. Environ Sci Technol 39:5268–5275

J.D. Istok, *Push-Pull Tests for Site Characterization*, Lecture Notes in Earth System Sciences 144, DOI 10.1007/978-3-642-13920-8,

Hageman KJ, Istok JD, Field JA, Buscheck TE, Semprini L (2001) In situ anaerobic transformation of trichlorofluoroethene in trichloroethene-contaminated groundwater. Environ Sci Technol 35(9):1729–1735

Hageman KJ, Field JA, Istok JD, Schroth MH (2003) "Forced mass balance" technique for estimating in situ transformation rates of sorbing solutes in groundwater. Environ Sci Technol 37:3920–3925

Haggerty R, Schroth MH, Istok JD (1998) Simplified method of "push-pull" test data analysis for determining in situ reaction rate coefficients. Ground Water 36(2):314–324

Haggerty R, McKenna SA, Meigs LC (2000) On the late-time behavior of tracer test breakthrough curves. Water Resour Res 36(12):3467–3479

Haggerty R, Fleming SW, Meigs LC, McKenna SA (2001) Tracer tests in a fractured dolomite. 2. Analysis of mass transfer in single-well injection-withdrawal tests. Water Resour Res 37(5):1129–1142

Hall SH, Luttrell SP, Cronin WE (1991) A method for estimating effective porosity and groundwater velocity. Ground Water 29(2):171–174

Hoopes JA, Harleman DRF (1967) Dispersion in radial flow from a recharge well. J Geophys Res 72(14):3595–3607

Istok JD, Humphrey MD, Schroth MH, Hyman MR, O'Reilly KT (1997) Single well, "push-pull" test for in situ determination of microbial activities. Ground Water 35(4):619–631

Istok JD, Field JA, Schroth MH, Sawyer TE, Humphrey MD (1999) Laboratory and field investigation of surfactant sorption using single-well, "push-pull" tests. Ground Water 37(4):589–598

Istok JD, Field JA, Schroth MH (2001) In situ determination of subsurface microbial enzyme kinetics. Ground Water 39(3):348–355

Istok JD, Field JA, Schroth MH, Davis BM, Dwarakanath V (2002) Single-well "push-pull" partitioning tracer test for NAPL detection in the subsurface. Environ Sci Technol 36:2708–2716

Istok JD, Senko JM, Krumholz LR, Bogle MA, Peacock A, Chang YJ, White DC (2004) In situ bioreduction of technetium and uranium in a nitrate-contaminated aquifer. Environ Sci Technol 38:468–475

Jury WA, Roth K (1990) Transfer functions and solute movement through soils: Theory and applications. Birkhauser, Basel, Switzerland

Kim Y, Istok JD, Semprini L (2004) Push-pull tests for assessing in situ aerobic cometabolism. Ground Water 42(3):329–337

Kim Y, Istok JD, Semprini L (2006) Push-pull tests evaluating in situ aerobic cometabolism of ethylene, propylene, and cis-dichloroethylene. J Contam Hydrol 82:165–181

Kleikemper J, Schroth MH, Sigler WV, Schmucki M, Bernasconi SM, Zeyer J (2002) Activity and diversity of sulfate-reducing bacteria in a petroleum hydrocarbon-contaminated aquifer. Appl Environ Microbiol 68:1516–1523

Leap DI, Kaplan PG (1988) A single-well tracing method for estimating regional advective velocity in a confined aquifer: theory and preliminary laboratory verification. Water Resour Res 24(7):993–998

Luthy L, Fritz M, Bachofen R (2000) In situ determination of sulfide turnover rates in a meromictic Alpine lake. Appl Environ Microbiol 66:712–717

McGuire JT, Long DT, Klug MJ, Haack SK, Hyndman DW (2002) Evaluating behavior of oxygen, nitrate, and sulfate during recharge and quantifying reduction rates in a contaminated aquifer. Environ Sci Technol 36:2693–2700

Meigs LC, Beauheim RL (2001) Tracer tests in a fractured dolomite .1. Experimental design and observed tracer recoveries. Water Resour Res 37(5):1113–1128

Mercado A (1966) Recharge and mining tests at Yavne 20 well field. Underground wate4r storage study tech. report 12. Publ. 611. Tahalp Water Plann for Isr., Tel Aviv

Pickens JF, Grisak GE (1981) Scale-dependent dispersion in a stratified granular aquifer. Water Resour Res 17(4):1191–1211

Pombo SA, Pelz O, Schroth MH, Zeyer J (2002) Field-scale 13C-labeling of phospholipid fatty acids (PLFA) and dissolved inorganic carbon: tracing acetate assimilation and mineralization in a petroleum hydrocarbon-contaminated aquifer. FEMS Microbiol Ecol 41:259–267

Pombo SA, Kleikemper J, Schroth MH, Zeyer J (2005) Field-scale Isotopic labeling of phospholipid fatty acids from acetate-degrading sulfate-reducing bacteria. FEMS Microbiol Ecol 51(2):197–207

Reinhard M, Shang S, Kitanidis PK, Orwin E, Hopkins GD, Lebron CA (1997) In situ BTEX biotransformation under enhanced nitrate- and sulfate- reducing conditions. Environ Sci Technol 31:28–36

Reusseur DE, Istok JD, Beller HR, Field JA (2002) In situ transformation of deuterated toluene and xylene to benzylsuccinic acid analogues in BTEX-contaminated aquifers. Environ Sci Technol 36:4127–4134

Schroth MH, Istok JD (2004) Approximate Solution for Solute Transport During Spherical Flow Push-Pull Tests. Ground Water 43(2):280–284

Schroth MH, Istok JD, Conner GT, Hyman MR, Haggerty R, O'Reilly KT (1998) Spatial variability in in situ aerobic respiration and denitrification rates in a petroleum-contaminated aquifer. Ground Water 36(6):924–937

Schroth MH, Istok JD, Haggerty R (2001) In situ evaluation of solute retardation using single-well push-pull tests. Adv Water Resour 24:105–117

Schürmann A, Schroth MH, Saurer M, Bernasconi SM, Zeyer J (2003) Nitrate-consuming processes in a petroleum-contaminated aquifer quantified using push-pull tests combined with 15N isotope and acetylene inhibition methods. J Contam Hydrol 66(1–2):59–77

Senko JM, Istok JD, Suflita JM, Krumholz LR (2002) In-situ evidence for uranium immobilization and remobilization. Environ Sci Technol 35:1491–1496

Snodgrass MF, Kitanidis PK (1998) A method to infer in situ reaction rates from push-pull experiments. Ground Water 36(4):645–650

Sternau R, Schwarz J, Mercado A, Harpaz Y, Nir A, Halevy E (1967) Radioisotope tracers in large scale recharge studies of ground water. Isotopes in Hydrology. Vienna, IAEA:489–506

Swartz CH, Gschwend PM (1999) Field studies of in-situ colloid mobilization in a southeastern coastal plain aquifer. Water Resour Res 35(7):2213–2223

Tomich JF, Dalton RL Jr, Deans HA, Shallenberger LK (1973) Single-well tracer method to measure residual oil saturation. J Petrol Technol 25:211–218

Trudell MR, Gillham RW, Cherry JA (1986) An in-situ study of the occurrence and rate of denitrification in a shallow unconfined sandy aquifer. Journal of Hydrology 83:251–268

Urmann K, Gonzalez-Gil G, Schroth MH, Hofer M, Zeyer J (2005) New field method: gas push-pull tests for the in-situ quantification of microbial activities in the vadose zone. Environ Sci Technol 39(1):304–310

Index

J.D. Istok, *Push-Pull Tests for Site Characterization*, Lecture Notes in Earth System Sciences 144, DOI 10.1007/978-3-642-13920-8,

Printed by Publishers' Graphics LLC
BT20121009.19.23.90